Workbook for
C N C
TECHNOLOGY
AND
PROGRAMMING

Steve Krar

Arthur Gill

Gregg Division
McGRAW-HILL PUBLISHING COMPANY

New York   Atlanta   Dallas   St. Louis   San Francisco
Auckland   Bogotá   Caracas   Hamburg   Lisbon
London   Madrid   Mexico   Milan   Montreal   New Delhi
Paris   San Juan   São Paulo   Singapore
Sydney   Tokyo   Toronto

Workbook for CNC: Technology and Programming

Copyright © 1990 by McGraw-Hill, Inc. All rights reserved. Printed in the United States of America. Except as permitted under the United States Copyright Act of 1976, no part of this publication may be reproduced or distributed in any form or by any means, or stored in a data base or retrieval system, without the prior written permission of the publisher.

1 2 3 4 5 6 7 8 9 0 SEMSEM 8 9 6 5 4 3 2 1 0 9

ISBN 0-07-023334-9

CONTENTS

To the Instructor	v
A Guide to the Use of This Workbook	vii
TEST ONE/History of Numerical Control	1
TEST TWO/History of Numerical Control	3
TEST THREE/The Computer	5
TEST FOUR/The Computer	7
TEST FIVE/Input Media	9
TEST SIX/Input Media	11
REVIEW TEST ONE	13
TEST SEVEN/How Numerical Control Operates Machine Tools	15
TEST EIGHT/How Numerical Control Operates Machine Tools	17
TEST NINE/Programming Data	19
TEST TEN/Programming Data	21
TEST ELEVEN/Simple Programming	23
TEST TWELVE/Simple Programming	25
REVIEW TEST TWO	27
TEST THIRTEEN/Angular and Contour Programming	29
TEST FOURTEEN/Angular and Contour Programming	31
TEST FIFTEEN/Miscellaneous Numerical Control Functions	33
TEST SIXTEEN/Miscellaneous Numerical Control Functions	35
TEST SEVENTEEN/Machining Centers	37
REVIEW TEST THREE	39
TEST EIGHTEEN/Chucking and Turning Centers	43
TEST NINETEEN/Electrical Discharge Machining	45
TEST TWENTY/Numerical Control and the Future	47

NC Milling

PROJECT ONE/Snoopy	49
PROJECT TWO/Horizontal Drill Block	50
PROJECT THREE/Vertical Drill Block	51
PROJECT FOUR/Drill Template 1	52
PROJECT FIVE/Drill Template 2	53
PROJECTS SIX AND SEVEN/Angular Slot Machining	54
PROJECT EIGHT/Base Plate	55
PROJECT NINE/Full Circle	56
PROJECT TEN/Two Arcs in Full Quadrants	57
PROJECT ELEVEN/Circular Groove Milling	58
PROJECT TWELVE/A Full Circle Through Arcs	59
PROJECT THIRTEEN/Bolt Circle Drilling	60

PROJECT FOURTEEN/Partial Circular Slot — 61
PROJECT FIFTEEN/Cutter Radius Tangency — 62
PROJECT SIXTEEN/Contouring Exercise — 63
PROJECT SEVENTEEN/Macro — 64
PROJECT EIGHTEEN/Alphabet — 65
PROJECT NINETEEN/Circular Alphabet — 66
PROJECT TWENTY/Cribbage Board — 67

NC Wire EDM

PROJECT ONE/Thread Center Gage — 69
PROJECTS TWO, THREE, FOUR, FIVE, AND SIX/Shapes 1–5 — 70
PROJECT SEVEN/C-Clamp — 73
PROJECT EIGHT/Sailboat — 74
PROJECT NINE/Witch — 75
PROJECT TEN/Violin — 76

NC Turning Center

PROJECT ONE/Two-Step Locating Pin — 77
PROJECT TWO/Taper Shaft — 78
PROJECT THREE/Spindle — 79
PROJECT FOUR/Contour Form — 80
PROJECT FIVE/Pawn — 81
PROJECT SIX/Bishop — 82
PROJECT SEVEN/Castle — 83
PROJECT EIGHT/Knight — 84
PROJECT NINE/Queen — 85
PROJECT TEN/King — 86

To the Instructor

This workbook is designed to be used with the text *CNC: Technology and Programming*. It contains two tests for each of Chapters 1 to 8, one each for Chapters 9 to 12, and three review tests. The questions are of the completion type, and each is designed to review a topic, stimulate students' thinking, and expand their knowledge of the subject matter. This workbook also contains a number of projects for machining centers, turning centers, and wire-cut EDM machines, which can be used as practical exercises or tests.

In the interests of fairness to the student and for ease of marking, it was the intention of the authors that each question would be of equal value. With this in mind, the value of each question and the time allotted for each test has been left to the discretion of the instructor. Generally, a test should take the student about three times longer than it would take an instructor to answer the same test.

The tests are arranged so that all the answers are placed on the right-hand side of the page. This format allows an easier method of checking student answers and a more uniform method of recording answers. In some cases, more than one word may be necessary or desirable to answer a question.

An instructor's manual for this book is available from the publisher. The wording of students' answers for some questions may vary somewhat from the authors'; therefore, the teacher should use discretion in deciding whether a student's answer is satisfactory and should mark it accordingly.

These tests should provide a good indication to the instructor as to the areas of difficulty encountered by each student. With this information, the instructor can then clarify any points which have been misunderstood.

The projects included in this workbook are for machining centers, turning centers, and EDM wire-cut machines. These projects are designed to introduce students to programming, starting with simple point-to-point positioning and progressing in a logical sequence to more complex programs. Programming sheets for these projects can be found in the Instructor's Manual (pages 30–34). Feel free to make as many copies as you need for class usage. Included are exercises in incremental and absolute programs for linear and circular interpolation, cutter radius tangencies, macros, drilling, milling, etc. Each project is designed to increase the student's programming skills and make him or her a suitable candidate for employment.

SPECIAL NOTES TO INSTRUCTORS

- Where only a computer plot of the program is given, the actual sizes can be scaled off the plotted drawings, digitized, or sized to suit each programmer.
- Program answers may vary, depending on the machine tool and NC controller available to the student.
- The program answers given are a guide for the instructor and may be changed to suit the equipment available.

The ways in which instructors use a text as a teaching and learning tool are bound to vary because of individual differences and background experiences. Since no one method of teaching suits everyone, teachers ought to use the method of instruction which works best for them. To be effective, however, an instructor should use as many teaching aids as possible, and actively involve students in the learning process.

S. F. KRAR
A. R. GILL

A Guide to the Use of This Workbook

The questions in this book are for use with the text *CNC: Technology and Programming*. Most tests are of the completion type, and all the answers are to be written in the Answer column at the right-hand side of each test page. To assist you in answering the questions and programming problems correctly, sample questions and answers are provided. Study these sample questions carefully so that you will follow the correct answering procedure.

TYPE 1 — COMPLETION TESTS

These questions contain spaces with a question mark __?__ and require a key word or words to make each sentence complete and true. All answers must be entered in the spaces provided in the right-hand column of the test page.

Example

		Answer
1. __?__ program locations are always given as the distance and direction from a single __?__ point.	1.	*absolute* *reference*
2. Two types of feedback systems for numerical control are __?__ and __?__.	2.	*digital* *analog*
3. The __?__ line is called the X axis and the __?__ line is called the Y axis.	3.	*horizontal* *vertical*

(Continued on next page)

TYPE 2 — PROJECT TESTS

Each project consists of a programming exercise where certain information is given and the student is required to complete the programming. See the following example and note that all programming steps are completed in the space provided in the right-hand column of the question page.

1. For the part shown:
   (a) Start at XY zero.
   (b) Include the preparatory function codes.
   (c) Program the part boundary *clockwise*, starting at point A in the absolute mode.
   (d) Return to XY zero.
   (e) Program the slot location, starting at point 1.
   (f) Return to XY zero.

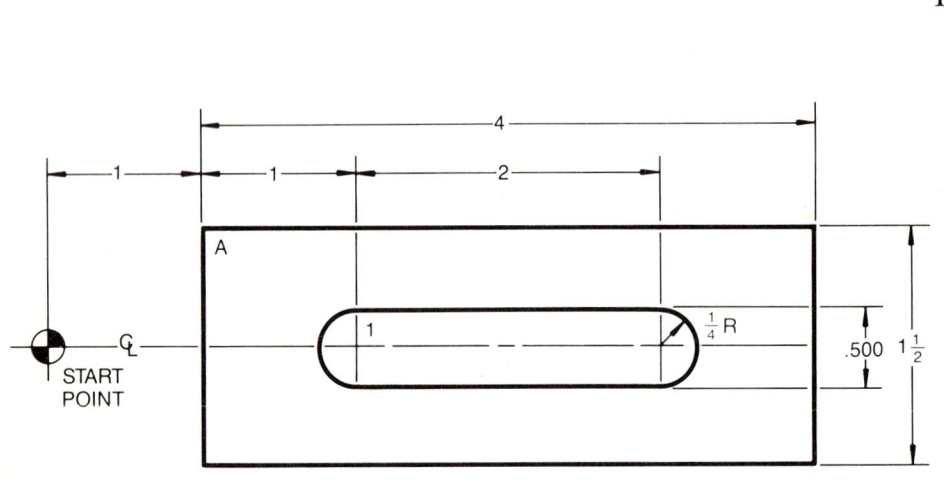

(The Superior Electric Company)

Answer

1. %
    N010 G90
    N020 G70
    N030 *G00 X1.000 Y0.750*
    N040 *G01 X5.000*
    N050 *Y−0.750*
    N060 *X1.000*
    N070 *Y0.750*
    N080 *G00 X0.0 Y0.0*
    N090 *X2.000 Y0.0*
    N100 *G01 X4.000*
    N110 *G00 X0.0 Y0.0*
    N120 M02
    %

viii

Name _____ Date _____

TEST ONE

History of Numerical Control

Study Chapter 1, History of Numerical Control, in your textbook.

Early forms of numerical control were used in 1725 on knitting machines, rotating drums to control chimes in European churches and cathedrals, and the punched paper rolls of player pianos. Today, numerical control is being used on all types of machine tools to produce complex parts automatically, to accurate dimensions, and reduce manufacturing costs.

Place the answers to the __?__ marks in each statement in the column at the right-hand side of the page.

Answer

1. The decimal system is based on the power of __?__, while the binary system is based on the power of __?__.

    1. _____
       _____

2. Numerical control data, which is processed by the machine control unit, consists of __?__, __?__, and __?__.

    2. _____
       _____
       _____

3. In a binary system, the number 1 means that the circuit is __?__, and the number 0 means that the circuit is __?__.

    3. _____
       _____

4. Chart 174 + 19 in binary numbers; be sure to complete the total (bottom) column.

    4.

DECIMAL		
100	10	1
1	7	4
+	1	9
1	9	3

(A)

BINARY							
256	128	64	32	16	8	4	2	1

(B)

5. In the coordinate system, a __?__ movement is always to the right of the zero or origin point, while a __?__ movement is to the left.

    5. _____
       _____

6. The __?__ and the __?__ programming systems are used in numerical control work.

    6. _____
       _____

7. __?__ program locations are always given as the distance and direction from a single __?__ point.

    7. _____
       _____

(Continued on next page)

Copyright © 1990 by McGraw-Hill, Inc. All rights reserved.   **History of Numerical Control** TEST ONE 1

8. Give the XY coordinate locations of each point on the diagram.

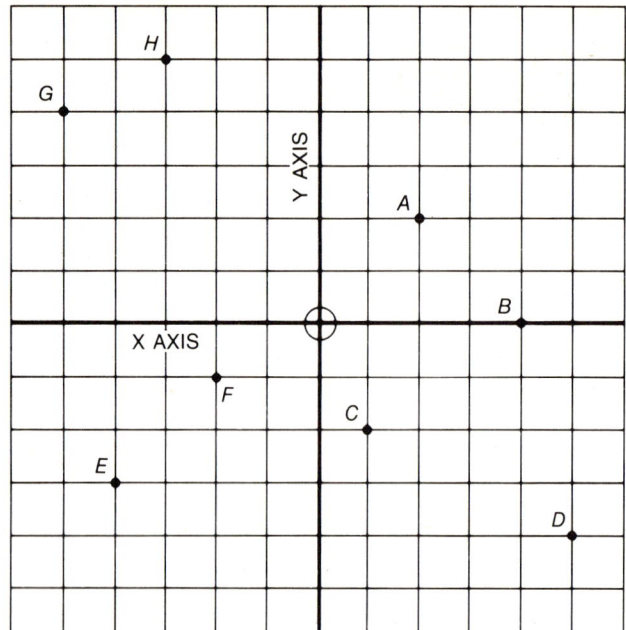

8. A _____
   B _____
   C _____
   D _____
   E _____
   F _____
   G _____
   H _____

9. The computer is only an extension of a person's brain and can perform tasks with amazing __?__ , __?__ , and __?__ .

9. _____
   _____
   _____

10. Three important sections or elements of a central processing unit are the __?__ unit, the __?__ unit, and the __?__ unit.

10. _____
    _____
    _____

11. Computer memory is generally classified as either __?__ or __?__ storage.

11. _____
    _____

TEST ONE History of Numerical Control

Name _____ Date _____

TEST TWO

History of Numerical Control

Study Chapter 1, History of Numerical Control, in your textbook.

Numerical control consists of using numbers, letters, and symbols to control a machine tool through a series of coded instructions which the machine control unit can understand. These coded instructions are converted into electrical pulses of current which the machine motors and controls follow to perform a machining operation on a workpiece.

Place the answers to the ? marks in each statement in the column at the right-hand side of the page.

Answer

1. The three developments in the electronics industry which spurred the use of numerical control in the machine tool industry were ? , ? , and ? .

   1. _____
      _____
      _____

2. Three early forms of numerical control were ? , ? , and ? .

   2. _____
      _____
      _____

3. Chart 365 + 25 in binary numbers; be sure to complete the total (bottom) column.

   3.

DECIMAL		
100	10	1
3	6	5
+	2	5
3	9	0

(A)

BINARY								
256	128	64	32	16	8	4	2	1

(B)

(Continued on next page)

4. Locate and label the following coordinate locations on the diagram.   4.

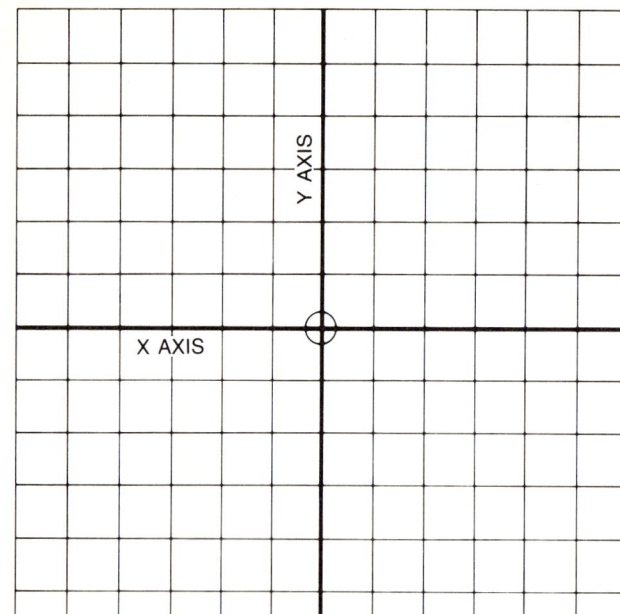

A = X3 Y-2
B = X-5 Y4
C = X3 Y2
D = X-2 Y-4
E = X-4 Y1
F = X-1 Y3
G = X4 Y-1
H = X1 Y-3

5. Most lathes are only programmed on the __?__ axis and the __?__ axis.   5. _____

6. Incremental program locations are always given as the __?__ and __?__ from the previous point.   6. _____

7. A digital computer accepts an __?__ of information, __?__ this information, and develops an __?__ data.   7. _____

8. The two types of machining centers are the __?__ and the __?__ type.   8. _____

9. __?__ - __?__ memory has a greater capacity and is more reliable than other forms of memory.   9. _____

10. The central processing unit of a computer contains three sections: the __?__ unit, the __?__ unit, and the __?__ unit.   10. _____

Name _____ Date _____

TEST THREE

The Computer

Study Chapter 2, The Computer, in your textbook.

Throughout recorded history, people have used some device to count and perform calculations. The world's first computer, the abacus, which involved moving beads on several wires, was developed in the Orient around 4000 B.C. This instrument may still be found used by some Oriental businesses to make fast and accurate calculations.

Place the answers to the __?__ marks in each statement in the column at the right-hand side of the page.

Answer

1. Five common types of input media used in numerical control are __?__ , __?__ , __?__ , __?__ , and __?__ .

2. The Electronic Industries Association (EIA) standard numerical control tape is __?__ in. wide, contains __?__ channels, and uses the __?__ - coded decimal system.

3. The # __?__ channel on an EIA tape is the odd-parity bit, while the # __?__ channel indicates the end of a block of information.

4. The two coding systems used for numerical control are the __?__ system and the __?__ system.

5. The two tape formats which have been standardized by the EIA are the __?__ and the __?__ formats.

6. The four stages of preparing a NC punched tape are __?__ , __?__ , __?__ , and __?__ .

7. The __?__ code refers to preparatory functions, while the __?__ code refers to miscellaneous functions.

8. A Z-axis motion *towards the zero point* or gage height would be a __?__ movement, while a motion *away from the zero point* would be a __?__ movement.

9. A programmer should always review the manuscript to check for __?__ , __?__ machine motion, and __?__ of unnecessary information.

Copyright © 1990 by McGraw-Hill, Inc. All rights reserved.

Name _____ Date _____

TEST FOUR

The Computer

Study Chapter 2, The Computer, in your textbook.

No other invention in the history of the world has had such a dramatic impact on society as the computer. Computers have revolutionized the efficient manufacture of all types of goods. In the overall manufacturing process, computers are used to design parts and to control machines during manufacture, testing, inspection, quality control, inventory control, and many other functions to ensure that parts are produced accurately and at the lowest cost.

Place the answers to the __?__ marks in each statement in the column at the right-hand side of the page.

Answer

1. Numerical control punched tapes are made of __?__ , __?__ , and __?__ .   1. _____

2. The five complete words or pieces of information contained in one block of numerical control information are __?__ , __?__ , __?__ , __?__ , and __?__ .   2. _____

3. A part can be programmed in either the __?__ or __?__ programming mode.   3. _____

4. The NC programmer must select the zero or origin point, which may be at any __?__ of the part, a point on the work-holding __?__ , or any point of the __?__ .   4. _____

5. A __?__ device can generate the __?__ path to see whether the correct part will be produced by the NC machine.   5. _____

6. What general information is contained in each word of this block of information?
   N040 G90 X3.125 Y0.750 M03   6. _____

7. Computer-aided design equipment is capable of producing a __?__ of the part and the __?__ to run NC machines in order to produce the part.   7. _____

8. A programmer should always review the manuscript to check for __?__ , __?__ , machine motion, and __?__ of unnecessary information.   8. _____

Copyright © 1990 by McGraw-Hill, Inc. All rights reserved.

Name _____ Date _____

TEST FIVE

Input Media

Study Chapter 3, Input Media, in your textbook.

The input media for numerical control are used to provide the control unit of the machine tool with the proper instructions to produce a desired part. Input media can be in several forms, such as manual, punched cards, magnetic tape, punched tape, diskette, or direct numerical control. Much of the new programming is done through direct numerical control (DNC).

Place the answers to the __?__ marks in each statement in the column at the right-hand side of the page.

Answer

1. Great advances in numerical control came as the result of new technology in the electronics industry, such as __?__, __?__, __?__, and __?__.

2. Three benefits of numerical control other than lower tooling costs and increased production are __?__, __?__, and __?__.

3. The purpose of the tape __?__ is to decode the __?__ on a punched tape and send it to the NC machine __?__ unit.

4. Two types of feedback systems for numerical control are __?__ and __?__.

5. Three sections of the central processing unit are __?__, __?__, and __?__.

6. Most CNC control units will automatically program in the absolute mode when the address code __?__ is used, and in the incremental mode when the code __?__ is used.

7. On soft-wired CNC units, it is the __?__ program in the computer __?__ which makes the control unit act or think like a turning center or a machining center.

8. Most machine control units can automatically compensate for variations in the cutter __?__ and __?__.

9. A subroutine or __?__, sometimes called a __?__ within a program, is used to store frequently used __?__ sequences which can be recalled from memory by a __?__.

Copyright © 1990 by McGraw-Hill, Inc. All rights reserved.

Input Media   TEST FIVE   9

Name _____ Date _____

TEST SIX

Input Media

Study Chapter 3, Input Media, in your textbook.

Numerical control consists of providing input media of coded instructions or commands which are listed in a logical sequence to have a machine tool perform a specific task or series of tasks in order to produce a finished part. Early forms of input media were manual, punched cards, or magnetic tape; new input media consist of punched tape, diskette, or direct numerical control.

Place the answers to the __?__ marks in each statement in the column at the right-hand side of the page.

Answer

1. The five key elements of a NC system which are involved from editing the punched tape to producing a finished part are __?__ , __?__ , __?__ , __?__ , and __?__ .

   1. _____
      _____
      _____
      _____
      _____

2. Four improvements which helped to make machine tools more reliable are __?__ , __?__ , __?__ , and __?__ .

   2. _____
      _____
      _____
      _____

3. Three factors which alone justify using numerical control are __?__ , __?__ , and __?__ .

   3. _____
      _____
      _____

4. __?__ devices are used to send data back to the machine control unit so that the machine __?__ position can be compared with the __?__ data.

   4. _____
      _____
      _____

5. An __?__ - __?__ system does not have a __?__ unit and therefore cannot compare the signals received from the machine __?__ unit with the amount the machine __?__ has moved.

   5. _____
      _____
      _____
      _____
      _____

6. The machine control unit takes binary-coded data from the punched tape and converts it into __?__ digits.

   6. _____

7. The main purpose of a machine control unit is to take the part program and __?__ this information into a __?__ that the machine tool can understand to __?__ the functions required to make a part.

   7. _____
      _____
      _____

8. __?__ conversion is the ability of a machine control unit to __?__ plus and minus (+ and −) values to produce a left-hand part from a right-hand program.

   8. _____

Copyright © 1990 by McGraw-Hill, Inc. All rights reserved.

Input Media TEST SIX 11

Name _____   Date _____

REVIEW TEST ONE

Study Chapters 1, 2, and 3, History of Numerical Control, The Computer, and Input Media, in your textbook.

Place the answers to the ? marks in each statement in the column at the right-hand side of the page.

                                                                Answer

1. The decimal system is based on the power of ? , while the binary system is based on the power of ? .
        1. _____
            _____

2. The three developments in the electronics industry which spurred the use of numerical control in the machine tool industry were ? , ? , and ? .
        2. _____
            _____
            _____

3. In the binary system, the number 1 means that the circuit is ? , and the number 0 means that the circuit is ? .
        3. _____
            _____

4. Chart 54 + 7 in binary numbers; be sure to complete the total (bottom) column.
        4.

DECIMAL

100	10	1
	5	4
+		7

(A)

BINARY

256	128	64	32	16	8	4	2	1

(B)

5. In the coordinate system, a ? movement is always to the right of the zero or origin point, while a ? movement is to the left.
        5. _____
            _____

(Continued on next page)

6. Locate and label the following coordinate locations on the diagram.   **6.**

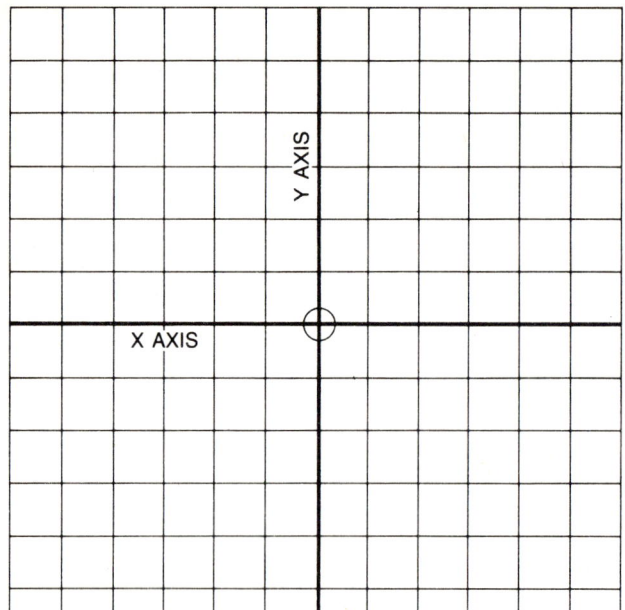

A = X2   Y-4
B = X-3 Y5
C = X-4 Y-1
D = X4   Y4

7. Three important sections or elements of a central processing unit are the __?__ unit, the __?__ unit, and the __?__ unit.   **7.** _____

8. The EIA standard numerical control tape is __?__ in. wide, contains __?__ channels, and uses the __?__-coded decimal system.   **8.** _____

9. The five complete words or pieces of information contained in one block of numerical control information are __?__, __?__, __?__, __?__, and __?__.   **9.** _____

10. The __?__ code refers to preparatory functions, while the __?__ code refers to miscellaneous functions.   **10.** _____

11. A part may be programmed in either the __?__ or __?__ programming mode.   **11.** _____

12. The purpose of the tape __?__ is to decode the __?__ on a punched tape and send it to the NC machine __?__ unit.   **12.** _____

13. The machine __?__ unit takes binary-coded data from the punched tape and converts it into __?__ digits.   **13.** _____

14. Most CNC control units will automatically program in the absolute mode when the address code __?__ is used, and in the incremental mode when the code __?__ is used.   **14.** _____

14   REVIEW TEST ONE   Copyright © 1990 by McGraw-Hill, Inc. All rights reserved.

Name _____ Date _____

TEST SEVEN

How Numerical Control Operates Machine Tools

Study Chapter 4, How Numerical Control Operates Machine Tools, in your textbook.

Numerical control is the process of automatically operating machine tools. This process takes coded information that is fed into the machine tool and converts it to an electrical form that the machine can understand. This information is then used to operate the machine, or it can be stored in the computer's memory and used at a later date.

Place the answers to the __?__ marks in each statement in the column at the right-hand side of the page.

Answer

1. The most common NC function codes are the __?__ or G functions, the __?__ or M functions, and the __?__ or A to W functions.

    1. _____
    _____
    _____

2. Sequence numbers are especially valuable when it becomes necessary to __?__ a tape and make __?__ to a program.

    2. _____
    _____

3. Preparatory functions can include such operations as __?__ -to- __?__ positioning, __?__ interpolation, and __?__ or __?__ programming.

    3. _____ -to- _____
    _____
    _____
    _____
    _____

4. The __?__ code is a drill cycle, while the __?__ code is a drill/dwell cycle.

    4. _____
    _____

5. Fixed or canned cycles, usually identified by the __?__ function codes G81 to G89, are a __?__ combination of operations which cause machine axis movement.

    5. _____
    _____

(Continued on next page)

6. For the part shown:
   (a) Start at XY zero.
   (b) Include the preparatory function codes.
   (c) Program the part boundary clockwise in the *incremental mode,* starting at point 1.
   (d) Return to XY zero.

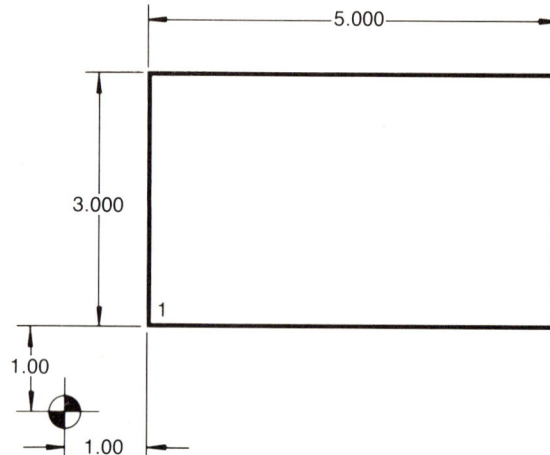

6.   %
N010 G91 (incremental)
N020 G70 (inch)
N030 _____
N040 _____
N050 _____
N060 _____
N070 _____
N080 _____
N090 %

7. For the part shown in Question 6:
   (a) Start at XY zero.
   (b) Include the preparatory functions.
   (c) Program the part boundary clockwise in the *absolute mode,* starting at point 1.
   (d) Return to XY zero.

7.   %
N010 G90 (absolute)
N020 G70 (inch)
N030 _____
N040 _____
N050 _____
N060 _____
N070 _____
N080 _____
N090 %

Name _____   Date _____

TEST EIGHT

How Numerical Control Operates Machine Tools

Study Chapter 4, How Numerical Control Operates Machine Tools, in your textbook.

As a result of new technology in the electronics industry, numerical control has made great advances since it was first introduced in the 1950s. From simple operations, the machine tools are now capable of doing complex machining operations with accuracy within a tolerance of 0.0001 to 0.0002 in. (0.0025 to 0.0050 mm).

Place the answers to the __?__ marks in each statement in the column at the right-hand side of the page.

Answer

1. The NC programmer's function is to take __?__ from an engineering drawing and __?__ it into __?__ that a machine control unit will understand.

    1. _____
       _____
       _____

2. The sequence number identifies each particular block of information; it usually consists of a __?__-digit number which is preceded by a __?__ code.

    2. _____
       _____

3. Identify the following preparatory function codes: (a) G00, (b) G70, (c) G90, (d) G96.

    3(a) _____
    (b) _____
    (c) _____
    (d) _____

4. Canned or fixed cycles can be used for __?__, __?__, and tapping operations.

    4. _____
       _____

5. Point-to-point positioning at a rapid rate is indicated by the __?__ code, while linear interpolation is indicated by the __?__ code.

    5. _____

(Continued on next page)

6. For the part shown:
   (a) Start at XY zero.
   (b) Include the preparatory function codes.
   (c) Program the part boundary clockwise in the *incremental mode,* starting at point 1.
   (d) Return to XY zero.

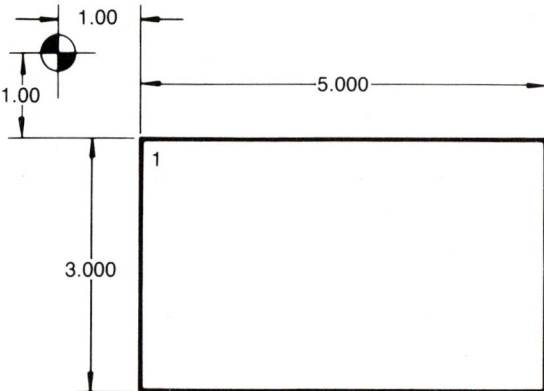

6.   %
   N010 G91 (incremental)
   N020 G70 (inch)
   N030 _____
   N040 _____
   N050 _____
   N060 _____
   N070 _____
   N080 _____
   N090 %

7. For the part shown in Question 6:
   (a) Start at XY zero.
   (b) Include the preparatory functions.
   (c) Program the part boundary clockwise in the *absolute mode,* starting at point 1.
   (d) Return to XY zero.

7.   %
   N010 G90 (absolute)
   N020 G70 (inch)
   N030 _____
   N040 _____
   N050 _____
   N060 _____
   N070 _____
   N080 _____
   N090 %

TEST NINE

Programming Data

Study Chapter 5, Programming Data, in your textbook.

The programmer's function is to take information from an engineering drawing and convert it into data that the control unit of a machine tool will understand. Various functions and codes are used in numerical control, and it is important that the programmer understand them so that the machine tool can be programmed to produce accurate parts. A knowledge of machining operations and procedures is very valuable to any programmer.

Place the answers to the ? marks in each statement in the column at the right-hand side of the page.

Answer

1. The three things that NC machine language consists of are __?__, __?__, and __?__.

2. The two distinct categories of numerical control programming systems are __?__ and __?__.

3. Point-to-point positioning is used to locate the machine __?__, or the __?__ at one or more specific __?__ to perform a machining operation.

4. During the point-to-point positioning between hole locations, the table moves at a __?__ rate along the __?__ axes on a __?__° angle line.

5. A __?__ or __?__ point should be established to permit the alignment of the workpiece and the machine tool.

6. The two most common forms of interpolation are __?__ and __?__.

(Continued on next page)

7. For the part shown:
   (a) Start at XY zero.
   (b) Include the preparatory function codes.
   (c) Program the part boundary clockwise in the *incremental mode*, starting at point 1.
   (d) Return to XY zero.
   (e) Program the location of the three holes.
   (f) Return to XY zero.

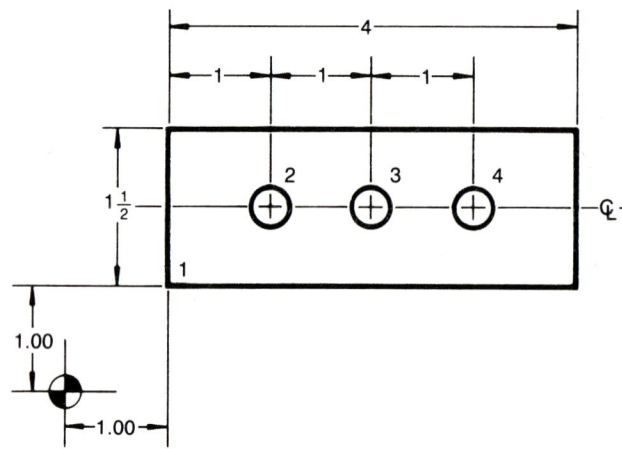

(The Superior Electric Company)

7.   %
N010 G91
N020 G70
N030 _____
N040 _____
N050 _____
N060 _____
N070 _____
N080 _____
N090 _____
N100 _____
N110 _____
N120 _____
N130 %

Name _____ Date _____

TEST TEN

Programming Data

Study Chapter 5, Programming Data, in your textbook.

Numerical control functions have been standardized by the EIA and the American Standard Code for Information Exchange (ASCII). Although there are slight variations between the systems, the function codes are basically the same. Preparatory and miscellaneous functions are used in numerical control work to indicate the type of operation that the machine tool is to perform.

Place the answers to the __?__ marks in each statement in the column at the right-hand side of the page.

Answer

1. Simple programming consists of taking __?__ from a part __?__ and converting this information into a __?__ that the machine tool understands.

2. Five methods of interpolation are __?__, __?__, __?__, __?__, and __?__.

3. __?__ interpolation consists of any programmed __?__ joined together by __?__ lines.

4. The four pieces of information required to program an arc by circular interpolation are __?__, __?__, __?__, and __?__.

(Continued on next page)

5. For the part shown:
   (a) Start at XY zero.
   (b) Include the preparatory function codes.
   (c) Program the part boundary clockwise in the *absolute mode,* starting at point 1.
   (d) Return to XY zero.
   (e) Program the location of the three holes.
   (f) Return to XY zero.

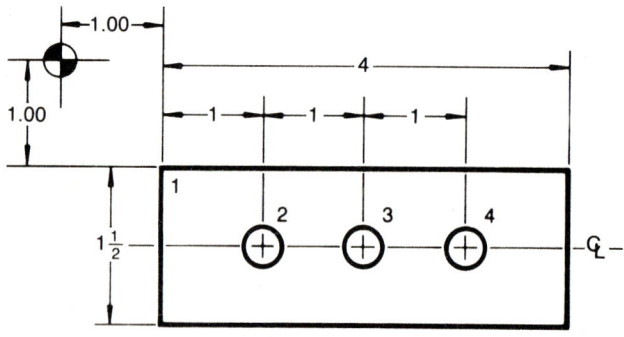

(The Superior Electric Company)

5. %
   N010 G90
   N020 G70
   N030 _____
   N040 _____
   N050 _____
   N060 _____
   N070 _____
   N080 _____
   N090 _____
   N100 _____
   N110 _____
   N120 _____
   N130 %

Name _____ Date _____

TEST ELEVEN
Simple Programming

Study Chapter 6, Simple Programming, in your textbook.

Simple programming involves taking the required information from a part drawing and converting this information into a language the machine tool can understand. The programmer should be familiar with machining operations and the tools required for each operation in order to produce the required part.

Place the answers to the __?__ marks in each statement in the column at the right-hand side of the page.

Answer

1. The rectangular coordinates __?__, __?__, and __?__ make it possible to exactly state the location of any point on a single plane.

   1. _____

2. Feed is generally measured in inches or millimeters per __?__ per __?__.

   2. _____

3. The word address letter __?__ refers to either the work surface or rapid-feed distance and is generally __?__ in. above the highest surface of the workplace.

   3. _____

4. __?__ zeros are those before the whole numbers to the __?__ of the decimal point.

   4. _____

5. The spindle speed rate is generally governed by three factors: __?__, __?__, and __?__.

   5. _____

(Continued on next page)

Copyright © 1990 by McGraw-Hill, Inc. All rights reserved.   **Simple Programming** TEST ELEVEN 23

6. For the part shown:
   (a) Start at XY zero.
   (b) Program in the *absolute mode*.
   (c) Program the part boundary clockwise and the hole locations in the numbered sequence.
   (d) Include sequence numbers, function codes (preparatory and miscellaneous), drill cycles, etc.
   (e) Return to XY zero.

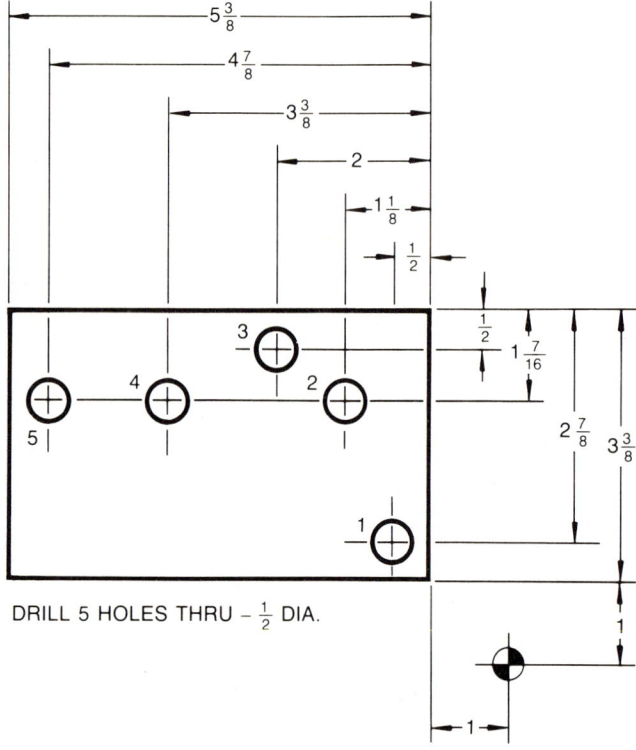

(The Superior Electric Company)

6. %
N010 G90
N020 G70

%

TEST ELEVEN Simple Programming

Name _____ Date _____

TEST TWELVE

Simple Programming

Study Chapter 6, Simple Programming, in your textbook.

In order to produce a simple program for any CNC machine, it is important that the programmer be familiar with the programming language, machining sequences, and tools required. The information from the part drawing is then used to produce a program that the machine control unit (MCU) for a particular machine can understand in order to produce the required part.

Place the answers to the __?__ marks in each statement in the column at the right-hand side of the page.

Answer

1. The __?__ line is called the X axis, the __?__ line is called the Y axis, and the point where they intersect is called the __?__ point.

   1. _____
      _____
      _____

2. Any X distance to the right of the Y axis is referred to as a __?__ dimension; X distances to the left are referred to as __?__ dimensions.

   2. _____
      _____

3. The R work plane is referred to as the __?__ dimension from which all programmed __?__ for cutting tools or machined surfaces are taken.

   3. _____
      _____

4. __?__ zeros are those after the last number to the __?__ of the decimal point.

   4. _____
      _____

5. Spindle speed is programmed in __?__ per minute and the word address letter __?__ followed by up to __?__ digits.

   5. _____
      _____
      _____

(Continued on next page)

Copyright © 1990 by McGraw-Hill, Inc. All rights reserved.

Simple Programming TEST TWELVE 25

6. For the part shown:
   (a) Start at XY zero.
   (b) Program in the *absolute mode*.
   (c) Program the part boundary *clockwise* and the hole locations in the numbered sequence.
   (d) Include sequence numbers, function codes (preparatory and miscellaneous), drill cycle, etc.
   (e) Return to XY zero.

6.   %
N010 G90
N020 G70

_____
_____
_____
_____
_____
_____
_____
_____
_____
_____
_____
_____
       %

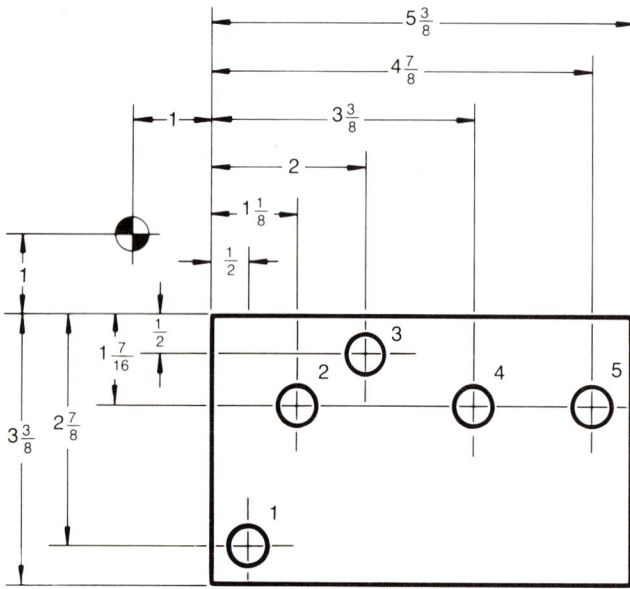

(The Superior Electric Company)

26 TEST TWELVE Simple Programming

Name _____ Date _____

REVIEW TEST TWO

Study Chapters 4, 5, and 6, How Numerical Control Operates Machine Tools, Programming Data, and Simple Programming, in your textbook.

Place the answers to the __?__ marks in each statement in the column at the right-hand side of the page.

                                                                                                                       Answer

1. The most common NC functions are the __?__ or G functions, the __?__ or M functions, and the __?__ or A to W functions.   1. _____

2. __?__ functions refer to some mode of __?__ of the machine tool or numerical control system.   2. _____

3. Sequence numbers are generally assigned in a progression of __?__ to leave enough room to insert as many as __?__ pieces of information before the program would have to be __?__.   3. _____

4. For the part shown:
   (a) Start at XY zero.
   (b) Include the preparatory functions.
   (c) Program the part boundary clockwise in the *incremental mode*, starting at point 1.
   (d) Return to XY zero.

    4.   %
    N010 G91
    N020 G70
    N030 _____
    N040 _____
    N050 _____
    N060 _____
    N070 _____
    N080 _____
    N090 %

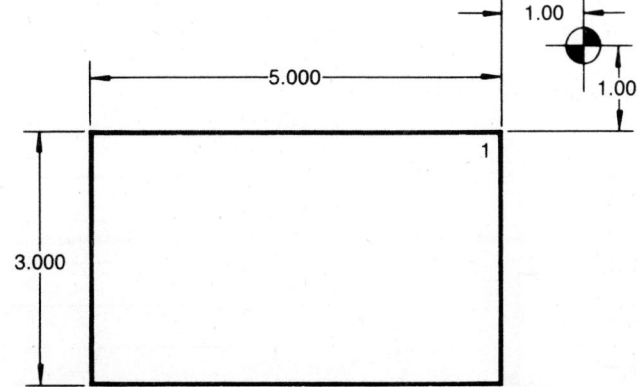

5. The three things that NC machine language consists of are __?__, __?__, and __?__.   5. _____

6. The four pieces of information required to program an arc by circular interpolation are __?__, __?__, __?__, and __?__.   6. _____

(Continued on next page)

7. For the part shown:
   (a) Start at XY zero.
   (b) Include all function codes.
   (c) Program the part boundary clockwise in the *absolute mode,* starting at point 1.
   (d) Return to XY zero.
   (e) Program the location of the three holes.
   (f) Return to XY zero.

7.
```
             %
N010 G90
N020 G70
N030 _____
N040 _____
N050 _____
N060 _____
N070 _____
N080 _____
N090 _____
N100 _____
N110 _____
N120 _____
             %
```

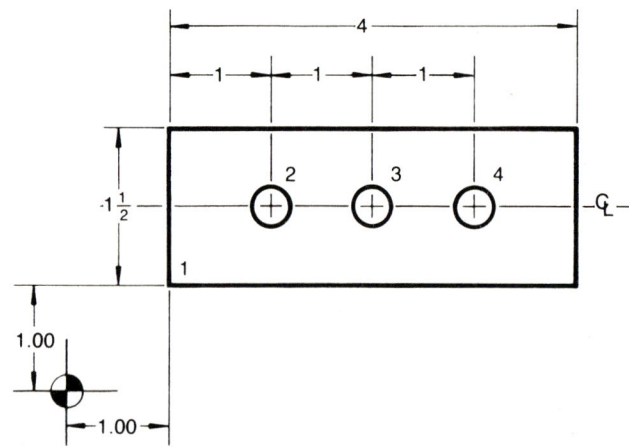

8. The __?__ line is called the X axis, the __?__ line is called the Y axis, and the point where they intersect is called the __?__ point.

8. _____
   _____
   _____

9. Spindle speed rate is generally governed by three factors: __?__, __?__, and __?__.

9. _____
   _____
   _____
   _____

10. For the part shown:
    (a) Start at XY zero.
    (b) Program in the *absolute mode.*
    (c) Program the part boundary *clockwise* and the hole locations in the numbered sequence.
    (d) Include sequence numbers, all function codes, drill cycle, etc.
    (e) Return to XY zero.

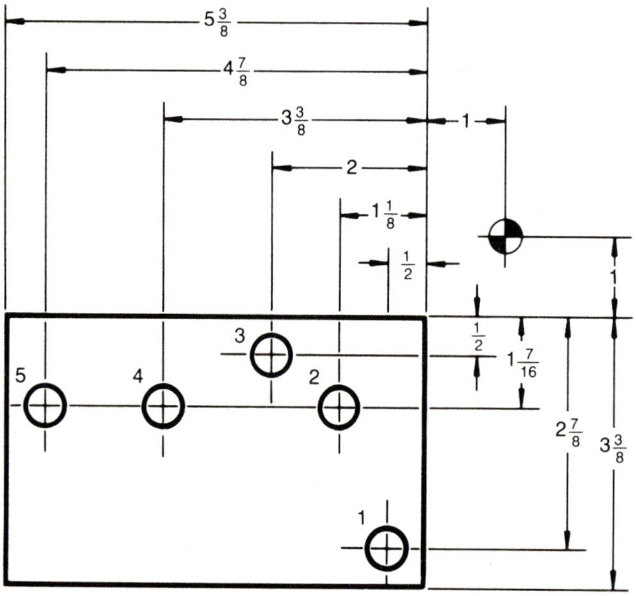

10.
```
             %
            G90
            G70
_____
_____
_____
_____
_____
_____
_____
_____
_____
_____
_____
_____
            M30
             %
```

DRILL 5 HOLES THRU – $\frac{1}{2}$ DIA.

28 REVIEW TEST TWO   Copyright © 1990 by McGraw-Hill, Inc. All rights reserved.

Name _____ Date _____

TEST THIRTEEN

Angular and Contour Programming

Study Chapter 7, Angular and Contour Programming, in your textbook.

Angular and contour programming can be achieved with the use of linear and circular interpolation. Since an angle is a straight line joining two points, the programmer needs to calculate only the coordinate locations of the start point and end point of the angle. Circles, arcs, and contours can be programmed using circular interpolation.

Place the answers to the __?__ marks in each statement in the column at the right-hand side of the page.

Answer

1. Whenever a part is to be machined with some type of milling cutter, the programmer must always calculate an __?__ path, which is usually __?__ the diameter of the cutter used.

    1. _____

2. The XY zero or home position of a machine is generally at the top __?__ corner of the table.

    2. _____

3. A machine control unit equipped with a __?__ zero allows the operator to fasten a workpiece anywhere on the machine table.

    3. _____

4. Some machine control units are equipped with __?__ - __?__ tool compensation to make allowances for __?__ in cutting tool length.

    4. _____

5. A programmer's tool list must include three pieces of information, such as __?__ , __?__ , and __?__ .

    5. _____
    _____
    _____

6. All cutting tools for a particular job are set against the same __?__ point, and the difference between each tool's actual length and the basic tool length is entered into memory as a __?__ value.

    6. _____
    _____

7. Newer MCUs automatically calculate cutter center-line offsets when the __?__ diameter is programmed.

    7. _____

(Continued on next page)

Copyright © 1990 by McGraw-Hill, Inc. All rights reserved.   **Angular and Contour Programming** TEST THIRTEEN **29**

8. For the part shown:
   (a) Begin at the start point (XY zero).
   (b) Program in the *absolute mode*.
   (c) Include all functions, sequence numbers, speeds, feeds, etc.
   (d) Material: ½-in.-thick mild steel plate (CS 100).
   (e) Cut the part boundary *clockwise* using a 1-in.-diameter high-speed steel four-flute end mill.
   (f) Cut the ½-in.-wide slot 0.250 in. deep with a ½-in.-diameter high-speed steel two-flute end mill, starting at point A.

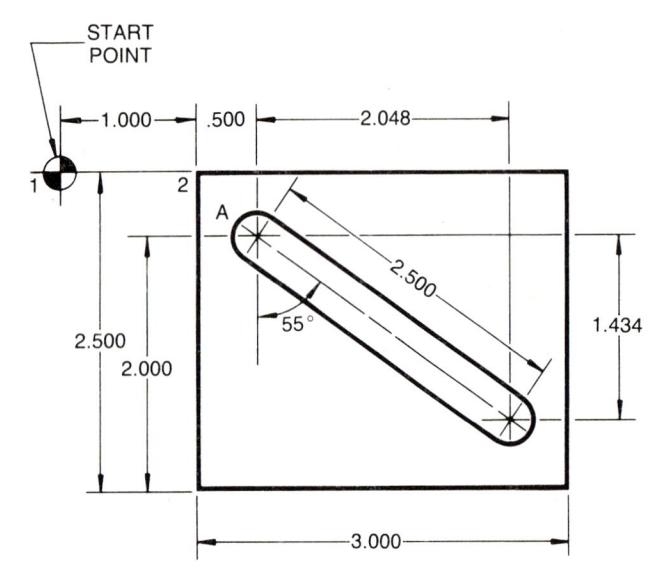

(The Superior Electric Company)

30 TEST THIRTEEN **Angular and Contour Programming**

Name _____ Date _____

TEST FOURTEEN

Angular and Contour Programming

Study Chapter 7, Angular and Contour Programming, in your textbook.

The programmer can program straight lines, angles, circles, arcs, and contours by using linear and circular interpolation. Once the coordinate locations have been calculated for the required angle or circle, the program is produced and stored in the MCU memory to be used to produce the required part.

Place the answers to the __?__ marks in each statement in the column at the right-hand side of the page.

                                                                         Answer

1. If a ¾-in.-diameter end mill is used, the programmer must keep __?__ offset from the normal surface to the edge of the __?__ which is doing the machining.

    1. _____

2. A full __?__ zero provides the operator with __?__ and reduces __?__ time because the XY zero can be located at any place on the machine table.

    2. _____

3. Three factors which are required to program an angle are __?__, __?__, and __?__.

    3. _____

4. When cutting angular surfaces tangent to a side or radius, it is necessary to calculate the cutter __?__ in relation to where the __?__ contacts the workpiece.

    4. _____

5. The amount of Z-axis travel is programmed from the __?__ plane, which is generally __?__ above the work surface.

    5. _____

(Continued on next page)

6. For the part shown:
   (a) Begin at the start point (XY zero).
   (b) Program in the *absolute mode*.
   (c) Include all functions, sequence numbers, speeds, feeds, etc.
   (d) Material: ½-in.-thick steel plate (CS 125).
   (e) Cut the part boundary *clockwise*, using a 1-in.-diameter high-speed steel four-flute end mill.
   (f) Cut the ½-in.-wide slot 0.250 in. deep with a ½-in.-diameter high-speed steel two-flute end mill, starting at point A.

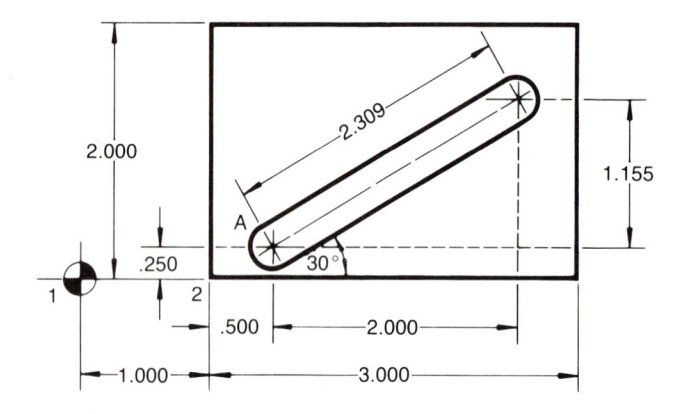

(The Superior Electric Company)

6. _____ %

32 TEST FOURTEEN **Angular and Contour Programming**

Name _____   Date _____

TEST FIFTEEN

Miscellaneous Numerical Control Functions

Study Chapter 8, Miscellaneous Numerical Control Functions, in your textbook.

Numerical control has made steady refinements over the years to the point where it is an invaluable and indispensable manufacturing tool. As technology improves, further advancements will be made to aid the programmer.

Place the answers to the ___?___ marks in each statement in the column at the right-hand side of the page.

                                                                                       Answer

1. Circular interpolation was developed to simplify the programming of ___?___ and ___?___ .

    1. _____

2. In circular interpolation, the I describes the ___?___ coordinate value, while the J describes the ___?___ coordinate value.

    2. _____

3. A parametric subroutine, sometimes called a ___?___ within a program, is used to store frequently used data ___?___ which can be recalled from memory when required.

    3. _____

4. The two most common methods of programming an arc are by ___?___ programming and by ___?___ programming.

    4. _____

5. The start point of an arc is usually the end point of a ___?___ line on the end point of a previous ___?___ .

    5. _____

6. Program the cutter path for the full circle shown:
    (a) Start at XY zero.
    (b) Program in a *clockwise* direction.
    (c) Use incremental programming.
    (d) Return to XY zero.

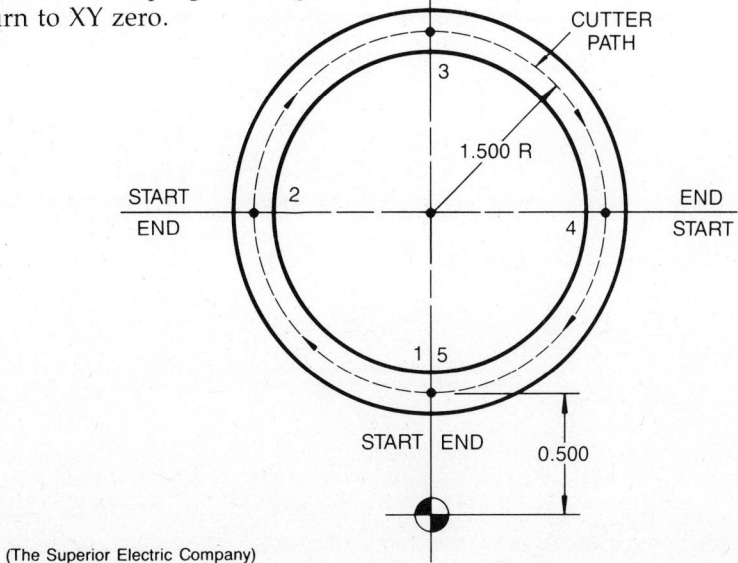

    6.   %
    N010 _____
    N020 _____
    N030 _____
    N040 _____
    N050 _____
    N060 _____
    N070 _____
    N080 _____
    N090 M02
           %

(The Superior Electric Company)

(Continued on next page)

Copyright © 1990 by McGraw-Hill, Inc. All rights reserved.   **Miscellaneous Numerical Control Functions**   TEST FIFTEEN **33**

7. Program the center path of the two arcs shown:
   (a) Start at XY zero.
   (b) Program the arcs in a *clockwise* direction, starting at point 1.
   (c) Use incremental programming.
   (d) Return to XY zero.

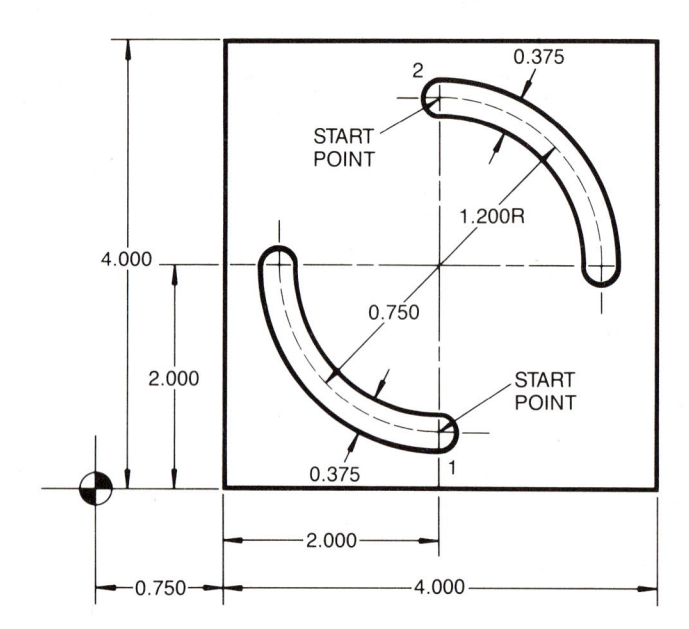

(The Superior Electric Company)

7.   %
N010 _____
N020 _____
N030 _____
N040 _____
N050 _____
N060 _____
N070 _____
N080 M02
     %

34   TEST FIFTEEN   **Miscellaneous Numerical Control Functions**   Copyright © 1990 by McGraw-Hill, Inc. All rights reserved.

Name _____   Date _____

TEST SIXTEEN

Miscellaneous Numerical Control Functions

Study Chapter 8, Miscellaneous Numerical Control Functions, in your textbook.

As computer technology advanced over the years since the early days of numerical control, the introduction of a large variety of computer languages has become available for NC programming. Although some of them are very specialized, it is advisable to learn a programming language that is popular and likely to be used for a long time.

Place the answers to the ? marks in each statement in the column at the right-hand side of the page.

                                                                                                      Answer

1. The four pieces of information required to program an arc or a circle are ? , ? , ? , and ? .

    1. _____
        _____
        _____
        _____

2. The circular interpolation of the MCU automatically breaks up an ? into very small ? moves to describe a circular path.

    2. _____
        _____

3. In circular interpolation, the center point of the arc is described by the letters ? and ? .

    3. _____
        _____

4. Programming a full circle requires ? start point(s) and ? end points.

    4. _____
        _____

5. Program the cutter path for the full circle shown:
   (a) Start at XY zero.
   (b) Program in a *clockwise* direction.
   (c) Use incremental programming.
   (d) Return to XY zero.

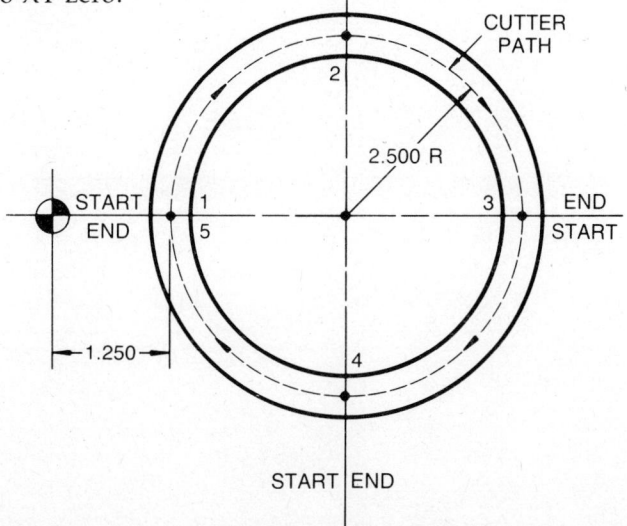

    5.   %
    N010 _____
    N020 _____
    N030 _____
    N040 _____
    N050 _____
    N060 _____
    N070 _____
    N080 _____
    N090 M02
              %

(Continued on next page)

6. Program the center path of the two arcs shown:
   (a) Start at XY zero.
   (b) Program the arcs in a *clockwise* direction, starting at point 1.
   (c) Use incremental programming.
   (d) Return to XY zero.

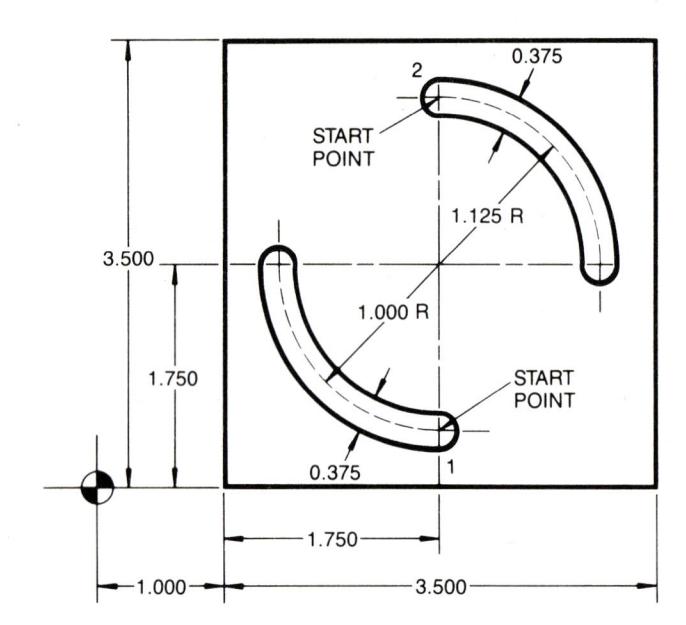

6.   %
N010 _____
N020 _____
N030 _____
N040 _____
N050 _____
N060 _____
N070 _____
N080 M02
         %

TEST SIXTEEN Miscellaneous Numerical Control Functions

Name _____ Date _____

TEST SEVENTEEN

Machining Centers

Study Chapter 9, Machining Centers, in your textbook.

Machining centers were developed in order to perform a number of different operations or machining sequences on a piece of work in one setup. Some machining centers are capable of operations such as milling, contouring, drilling, counterboring, boring, spot-facing, and boring on a part in any sequence while requiring only one setup of the workpiece.

Place the answers to the ? marks in each statement in the column at the right-hand side of the page.

                                                                                                      Answer

1. The two main types of machining centers are the ? spindle and the ? spindle machines.

    1. _____

2. The ? -column and the ? -column machining centers are two types of ? spindle machines.

    2. _____

3. The servo system consists of ? - ? motors, ? screws, and position ? encoders to provide fast, accurate movement and positioning of the XYZ axes slides.

    3. _____

4. The two most common types of tool changers are the ? tool changer and the ? tool changer.

    4. _____

5. Numerical control ? are used to accurately ? a part and ? it securely for any machining operation.

    5. _____

6. The most important factors which affect the efficiency of a machining operation are ? , ? , and ? .

    6. _____

7. The three most common methods of identifying machining center tools are ? , ? , and ? .

    7. _____

8. ? tooling selection refers to a system where there is no specific pattern of tool selection, while ? tooling selection refers to a system where all tools are loaded in the exact order in which they are used.

    8. _____

9. ? control machining increases productivity by sensing or ? damage to the cutting tool by reducing the feed rate, turning on the coolant, or stopping the machining cycle.

    9. _____

10. G codes in groups 1 to 10 are ? and do not have to be repeated in every block, while those in group 00 are ? and have to be repeated in every block in which they are used.

    10. _____

Copyright © 1990 by McGraw-Hill, Inc. All rights reserved.

Name _____ Date _____

REVIEW TEST THREE

Study Chapters 7, 8, and 9, Angular and Contour Programming, Miscellaneous Numerical Control Functions, and Machining Centers, in your textbook.

Place the answers to the ? marks in each statement in the column at the right-hand side of the page.

                                                                                 Answer

1. If a ¾-in.-diameter end mill is used, the programmer must keep   ?   offset from the normal surface to the edge of the   ?   which is doing the machining.

    1. _____

2. A tool assembly drawing should   ?   each cutting tool and give its   ?   length.

    2. _____

3. When cutting angular surfaces which are tangent to another surface or radius, the   ?   of the cutter is programmed.

    3. _____

4. A machine control unit equipped with a   ?   zero allows the operator to fasten a workpiece anywhere on the machine table.

    4. _____

5. When cutting angular surfaces to a side or radius, it is necessary to calculate the cutter   ?   in relation to where the   ?   contacts the workpiece.

    5. _____

6. For the part shown:
   (a) Begin at the start point (XY zero).
   (b) Program in the absolute mode.
   (c) Include all functions, sequence numbers, speeds, feeds, etc.
   (d) Material: mild steel ½ in. thick (CS 100).
   (e) Cut the part boundary *clockwise*, using a 1-in.-diameter HSS four-flute end mill.
   (f) Cut the ½-in.-wide slot, 0.250 in. deep, starting at point A.

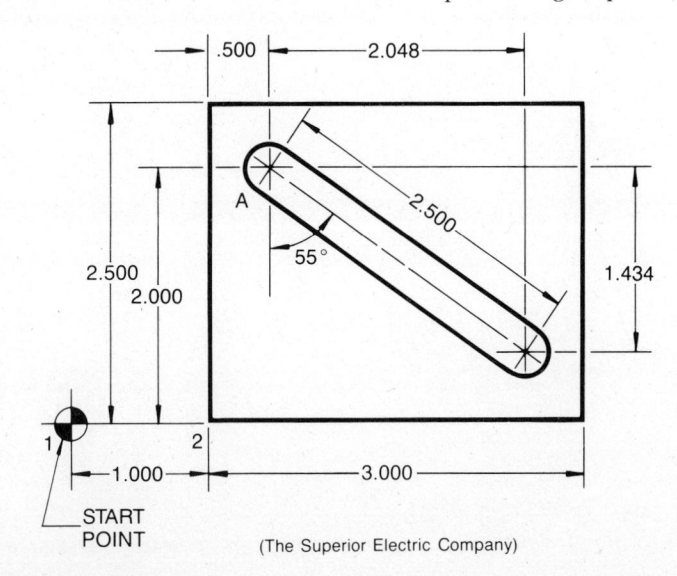

    6.   %
       N010 G90

       _____
       _____
       _____
       _____
       _____
       _____
       _____
       _____
       _____
       _____
       _____
       _____

       M30
       %

(The Superior Electric Company)

(Continued on next page)

7. Circular interpolation was developed to simplify the programming of __?__ and __?__ .

8. The program code for circular interpolation clockwise is __?__ ; that for circular interpolation counterclockwise is __?__ .

9. The two most common methods of programming an arc are by __?__ programming and by __?__ programming.

10. A parametric subroutine, sometimes called a __?__ within a program, is used to store frequently used data __?__ which can be recalled from memory as required.

11. Program the cutter path for the full circle:
    (a) Start at XY zero.
    (b) Program clockwise in the incremental mode.
    (c) Return to XY zero.

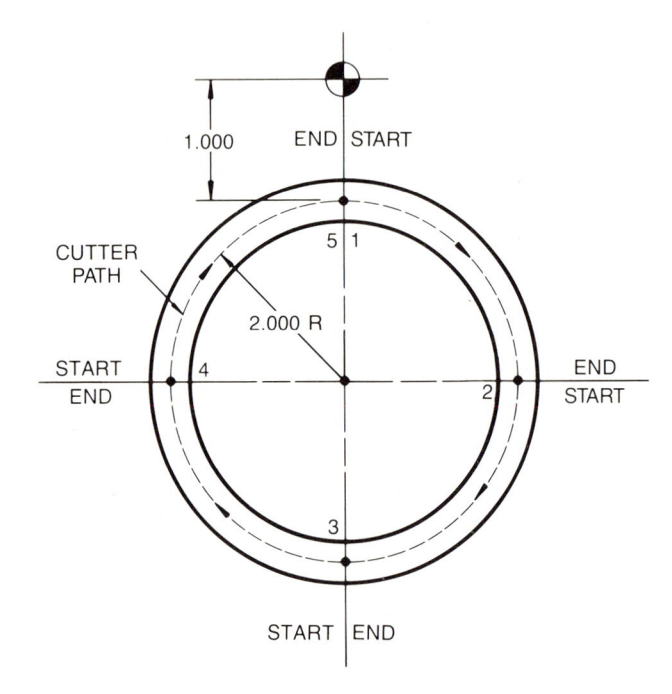

(The Superior Electric Company)

7. _____

8. _____

9. _____

10. _____

11.   %
N010 G91
N020 _____
N030 _____
N040 _____
N050 _____
N060 _____
N070 _____
N080 _____
N090 M30
            %

Name _____ Date _____

12. Program the center path of the two arcs:
    (a) Start at XY zero.
    (b) Program in the incremental mode.
    (c) Program the arcs clockwise, starting at point 1.
    (d) Return to XY zero.

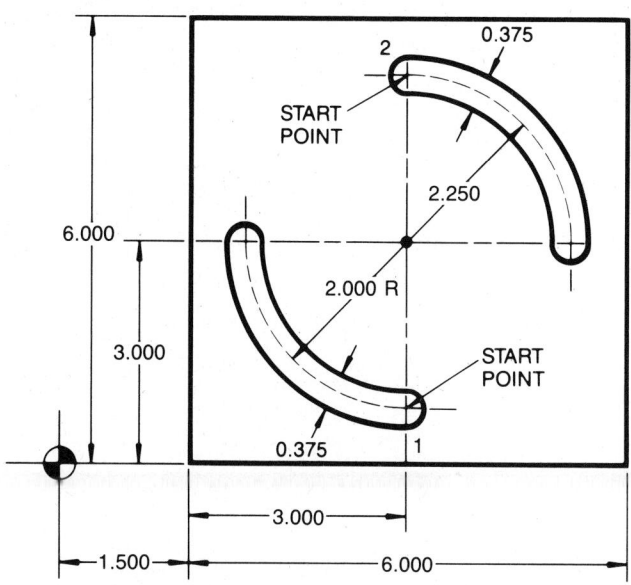

12.   %
N010 G91
N020 G70
N030 _____
N040 _____
N050 _____
N060 _____
N070 _____
N080 M30
        %

(The Superior Electric Company)

13. The two main types of machining centers are the __?__ spindle and the __?__ spindle machines.

13. _____

14. The servo system consists of __?__-__?__ motors, __?__ screws, and position __?__ encoders to provide fast, accurate movement and positioning of the XYZ axes.

14. _____

15. The most important factors which affect the efficiency of a machining operation are __?__, __?__, and __?__.

15. _____

16. __?__ tooling selection refers to a system where there is no specific pattern of tool selection, while __?__ tooling selection refers to a system where all tools are loaded in the exact order in which they are used.

16. _____

Copyright © 1990 by McGraw-Hill, Inc. All rights reserved.   **REVIEW TEST THREE**   41

Name _____ Date _____

TEST EIGHTEEN

Chucking and Turning Centers

Study Chapter 10, Chucking and Turning Centers, in your textbook.

Numerical control chucking and turning centers are capable of greater precision and higher production rates than were possible with the standard lathes they are replacing. These machine tools can be equipped with tool changers, automatic part loading and unloading devices, in-process gaging, and machining monitors, and can run virtually unattended with a minimum amount of downtime. CNC chucking and turning centers are very versatile machine tools and are cost-effective on production runs.

Place the answers to the __?__ marks in each statement in the column at the right-hand side of the page.

Answer

1. The four-axis chucking center incorporates __?__ turrets operating independently on __?__ slides to machine the workpiece simultaneously.

2. On a four-axis chucking center, the upper turret can machine the __?__ diameter, while the lower turret machines the __?__ diameter.

3. When longer parts are machined, the right-hand end of the shaft may be supported with a __?__ mounted in the __?__ turret.

4. Turning centers, while not unlike chucking centers, are designed mainly for machining __?__ - __?__ workpieces.

5. When a turning center is equipped with a steadyrest, operations such as __?__ and __?__ may be performed on the end of the shaft.

6. Center-drive lathes make it possible to machine parts from __?__ sides without relocating the workpiece.

7. Two devices can be used to measure the clamping force of a chuck; one measures __?__ gripping force, and one measures __?__ gripping force.

8. Always use the maximum chuck clamping pressure unless the pressure applied will __?__ the part.

9. Centrifugal force __?__ as the speed of rotation increases.

10. When machining long, thin workpieces, __?__ and __?__ can be minimized by using a steadyrest.

11. In-process gaging systems are a way of __?__ what is happening to the __?__ and the __?__ during the machining operations.

(Continued on next page)

Copyright © 1990 by McGraw-Hill, Inc. All rights reserved.   **Chucking and Turning Centers** TEST EIGHTEEN 43

12. The use of in-process gaging helps to __?__ operator errors in setup, and allows for __?__ of fully machined parts in the machine.

12. _____
    _____

13. On a chucking or turning center, the X plus (+) movement moves the cutting tool __?__ the center line, and the Z plus (+) movement moves the cutting tool __?__ the headstock.

13. _____
    _____

14. The most common way to detect tool wear is by the __?__ or __?__ it takes to drive the cutting tool.

14. _____
    _____

Name _____ Date _____

TEST NINETEEN

Electrical Discharge Machining

Study Chapter 11, Electrical Discharge Machining, in your textbook.

Electrical discharge machining (EDM) is the process of removing metal with an electrical discharge. This process is especially useful for machining very hard materials or parts with very thin, deep sections. Electrical discharge machining has been widely accepted in the tool and die industry, and is continuing to expand as more and more applications are being found for this process.

Place the answers to the __?__ marks in each statement in the column at the right-hand side of the page.

Answer

1. Electrical discharge machining removes metal by using a __?__ duration and high __?__ density electrical discharge between the __?__ and the __?__.

    1. _____
    _____
    _____
    _____

2. There are two types of EDM machines used in industry, the __?__ EDM machine and the __?__ EDM machine.

    2. _____
    _____

3. The __?__ fluid carries away the eroded particles of metal and also helps to dissipate the __?__ created by the spark.

    3. _____
    _____

4. Electrical discharge machines are equipped with a __?__ control mechanism that __?__ maintains a constant gap between the wire and the workpiece.

    4. _____
    _____

5. If the gap is too large, __?__ of the dielectric fluid does not occur, and machining __?__ take place.

    5. _____
    _____

6. If the gap is too small, the wire will touch the workpiece, causing it to __?__ and __?__.

    6. _____
    _____

7. Poor flushing conditions will cause __?__ cutting and __?__ machining conditions.

    7. _____
    _____

8. The amount of __?__ is measured by its specific resistance, the __?__ the resistance, the __?__ the cutting speed.

    8. _____
    _____
    _____

9. If __?__ sparks occur during the cutting operation, the water supply is __?__.

    9. _____
    _____

(Continued on next page)

10. If the wire electrode breaks during the machining cycle, the machine will enter into a __?__ condition.

11. Taper cutting is performed by the simultaneous control of the __?__ wire guide and the __?__ wire guide.

12. __?__ cutting on the wire-cut EDM is similar to taking a finish cut with a conventional machine tool.

10. _____

11. _____
    _____

12. _____

Name _____ Date _____

TEST TWENTY

Numerical Control and the Future

Study Chapter 12, Numerical Control and the Future, in your textbook.

Initially, numerical control programs were manually prepared; however, because this was too slow and cumbersome for complex parts, various programming languages were developed to make this job easier. Today, NC programming is assisted by computer-aided design (CAD), where the shape of a complex part can be translated into a NC program. It is quite likely that numerical control systems will play an important role in joining engineering and manufacturing information for the factory of the future.

Place the answers to the __?__ marks in each statement in the column at the right-hand side of the page.

Answer

1. In the factory of the future, the entire operation of a manufacturing plant will be __?__ and __?__ by a communication network that joins __?__ data design with __?__ data.

2. In DNC, a computer is linked directly to the machine __?__ unit and can load __?__ for many machine tools from a central __?__ .

3. Most CAD systems consist of a __?__ , __?__ , __?__ , and __?__ which help an operator design a part on the screen.

4. Four advantages of a CAD system are __?__ , __?__ , __?__ , and __?__ .

5. Computer-aided manufacturing (CAM) ties together all major functions of a factory, such as __?__ , __?__ , __?__ , __?__ , __?__ , and __?__ .

6. A CAM system generally contains three major divisions, __?__ , __?__ , and __?__ .

7. The future of the automated and unattended factory is dependent on how soon the manufacturing __?__ programs will become available.

Copyright © 1990 by McGraw-Hill, Inc. All rights reserved. Numerical Control and the Future TEST TWENTY 47

NC Milling

PROJECT ONE

Snoopy

This project involves the programming of the part boundary of the piece and the location of the three holes.

Notes

1. Use incremental programming.
2. Start at XY zero (located at left of part).
3. Program the part boundary *clockwise,* starting at point A.
4. Return to XY zero (start position).
5. Program the three hole locations in the numbered sequence.
6. Return to XY zero.

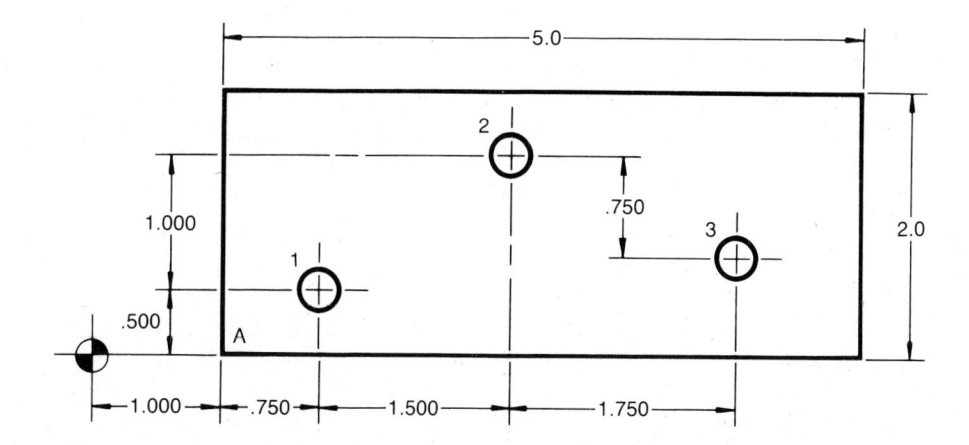

PROJECT TWO

Horizontal Drill Block

This project involves the programming of the part boundary of the piece and the location of the three holes along the X axis in the incremental programming and absolute programming modes.

Notes

1. Two programs are required:
   (a) Using incremental programming.
   (b) Using absolute programming.
2. Start at XY zero (located at the left of the part).
3. Program the part boundary *clockwise*, starting at point A.
4. Return to XY zero (start position).
5. Program the three hole locations in the numbered sequence.
6. Return to XY zero.

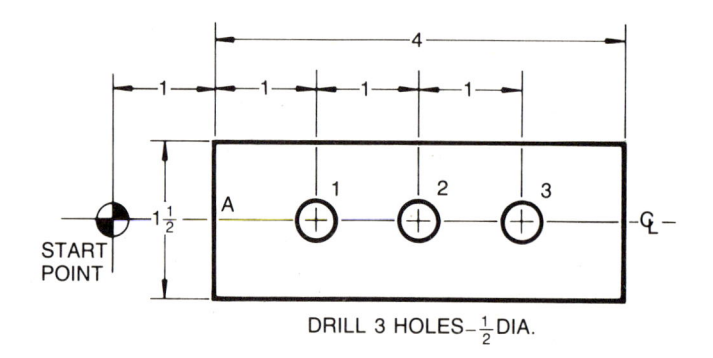

DRILL 3 HOLES—$\frac{1}{2}$ DIA.

(The Superior Electric Company)

PROJECT THREE

Vertical Drill Block

This project involves the programming of the part boundary of the piece and the location of the three holes along the X axis, using incremental programming and absolute programming.

Notes

1. Two programs are required:
   (a) Using incremental programming.
   (b) Using absolute programming.
2. Start at XY zero (located at the left of the part).
3. Program the part boundary *clockwise,* starting at point A.
4. Return to XY zero (start position).
5. Program the three hole locations in the numbered sequence.
6. Return to XY zero.

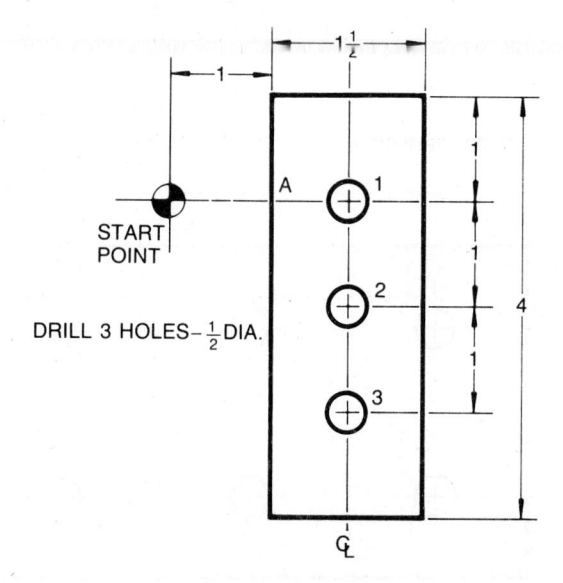

(The Superior Electric Company)

PROJECT FOUR

Drill Template 1

This project involves the programming of the part boundary of the piece and the drilling of the six holes, using the Z-axis motion for the drilling operation.

Notes

1. Two programs are required:
   (a) Using incremental programming.
   (b) Using absolute programming.
2. Start at XY zero.
3. Program the part boundary *clockwise,* starting at point A.
4. Return to XY zero.
5. Use the Z-axis motion to drill.
6. Drill the six holes in the numbered sequence.
6. Return to XY zero.

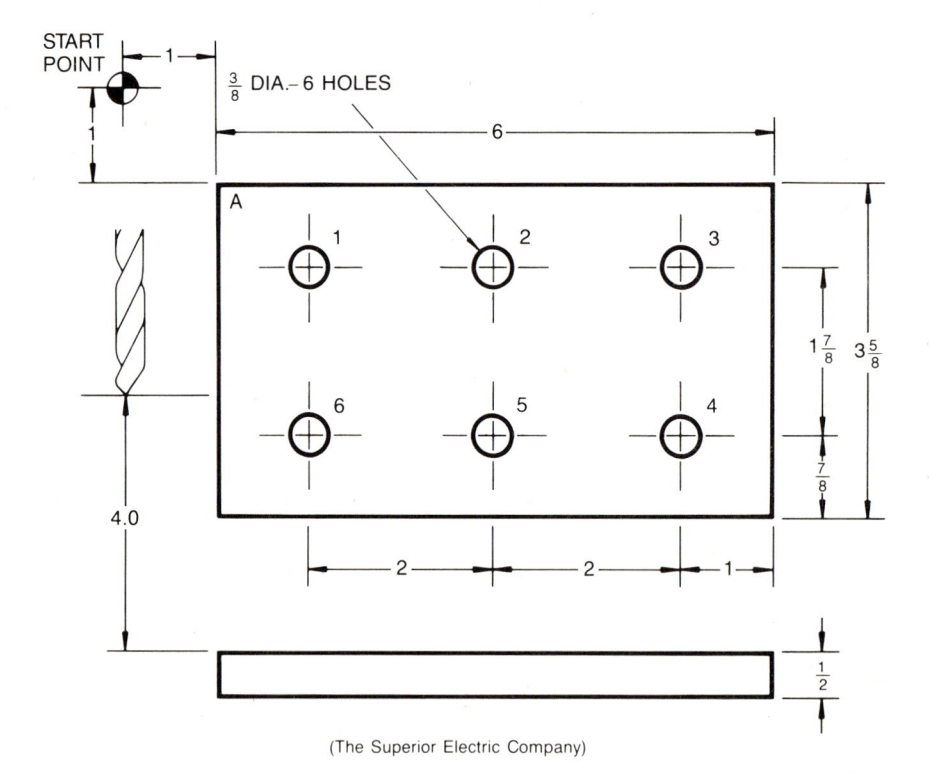

(The Superior Electric Company)

PROJECT FIVE

Drill Template 2

This project involves the milling (programming the cutter path) of the part boundary, center drilling, and drilling the holes, using the G81 fixed cycle for drilling similar holes in sequence. Proper speeds and feeds should be calculated and programmed for the type of material being cut.

Notes

1. Material: aluminum ½ in. thick (CS 300).

2. High-speed steel cutting tools are being used:
   (a) 1-in.-diameter four-flute end mill.
   (b) ½-in.-diameter drill.

3. Program *clockwise* in the absolute mode.

4. Include all speeds, feeds, and codes required to machine this part.

5. Use coolant when machining.

6. Use the G81 fixed cycle for center drilling and drilling the holes.

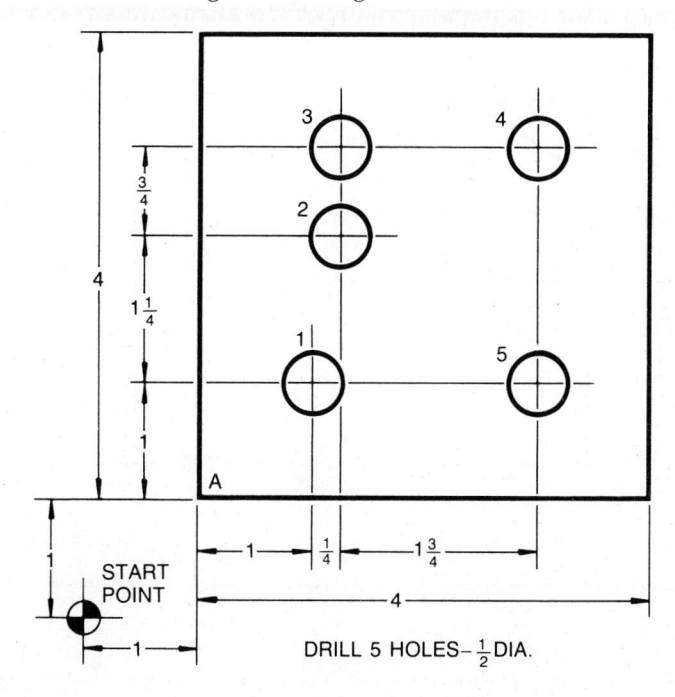

DRILL 5 HOLES – ½ DIA.

(The Superior Electric Company)

PROJECTS SIX AND SEVEN

Angular Slot Machining

These projects involve the milling of the part boundary and the angular slot. The operations involve the use of rapid traverse, tool feeds, and retract codes. The following notes should be carefully followed to ensure proper programming and machining of the parts.

Notes

1. Material: aluminum ½ in. thick (CS 300).

2. High-speed steel cutting tools are being used:
   (a) 1-in.-diameter four-flute end mill (for sides).
   (b) ¼-in.-diameter two-flute end mill (for slot).

3. Program clockwise in the absolute mode.

4. Start milling the slot at point A.

5. Include all speeds, feeds, and codes required to machine these parts.

6. Use coolant when machining.

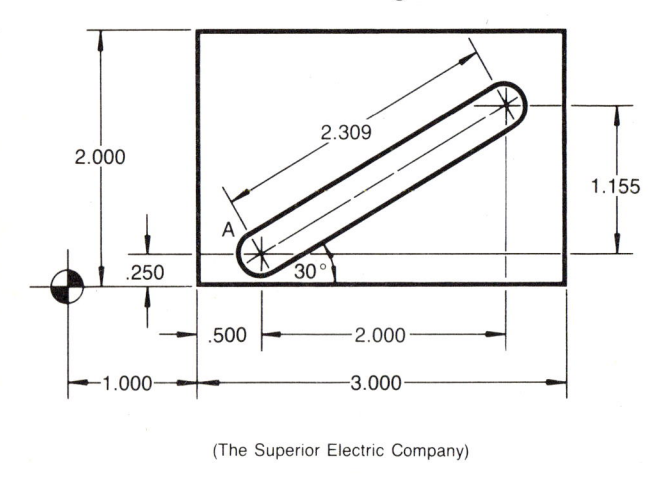

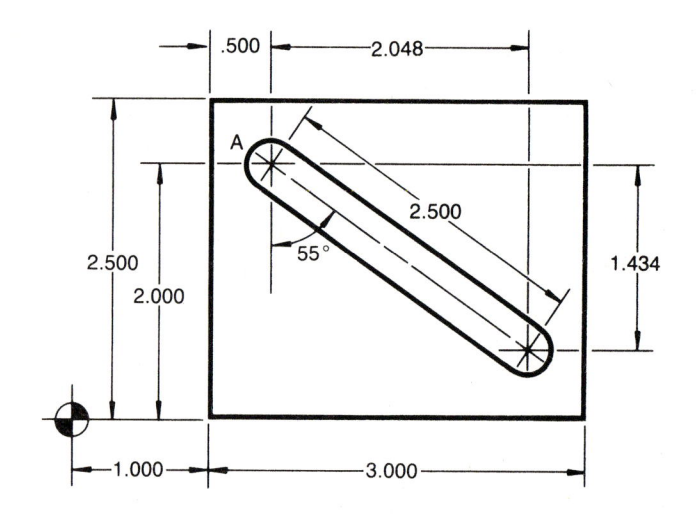

(The Superior Electric Company)

(The Superior Electric Company)

PROJECT EIGHT

Base Plate

This project involves the milling of the part boundary and the three slots in the base plate. It consists of programming for milling when more than one tool feed and retract cycle are necessary. Calculate proper speeds and feeds, and include all codes required to machine the part.

Notes

1. Material: aluminum (CS 250).
2. High-speed steel cutting tools are being used:
   (a) 1-in.-diameter four-flute end mill.
   (b) ¼-in.-diameter two-flute end mill.
3. Program in absolute mode.
4. Mill the part boundary clockwise, and the slots in numbered sequence.
5. Include all speeds, feeds, and codes required.

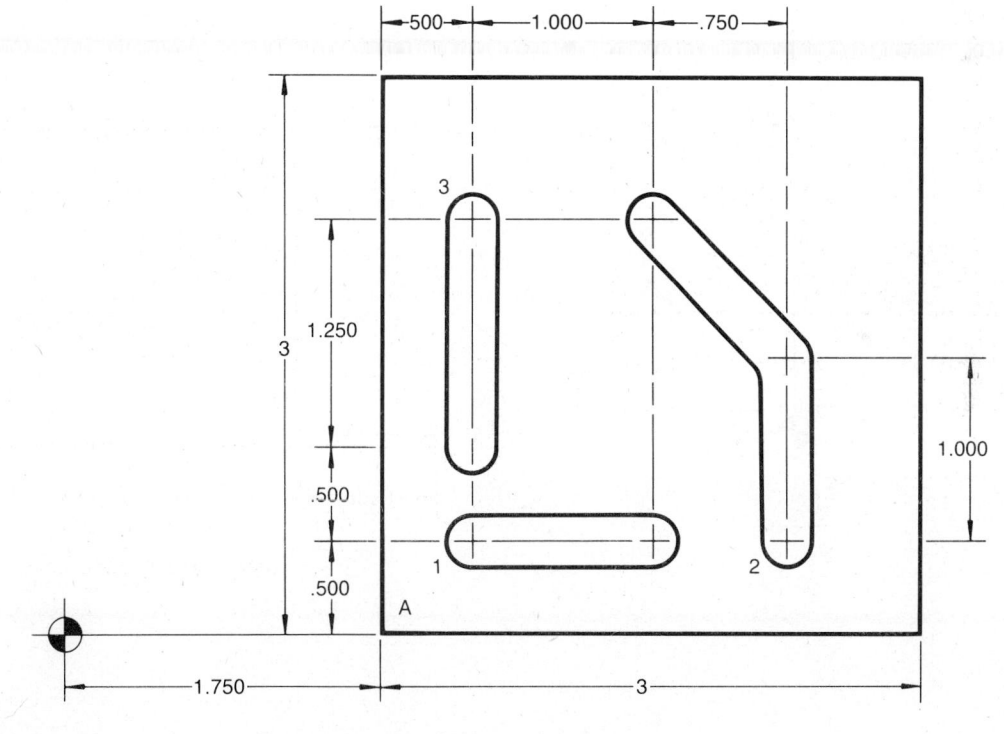

(The Superior Electric Company)

NC Milling PROJECT EIGHT

PROJECT NINE

Full Circle

This project involves the use of circular interpolation to program the cutter path for a full circle. The programmer can make the cutting tool follow any circular path ranging from a small arc segment to a full circle. The use of the I and J codes is required to locate the center point of the radius or arc.

Notes

1. The diameter of the circle is 3 in.
2. Start at XY zero.
3. Program the part cutter path starting at point A:
   (a) Clockwise in the absolute mode.
   (b) Counterclockwise in the incremental mode.
4. Return to XY zero.

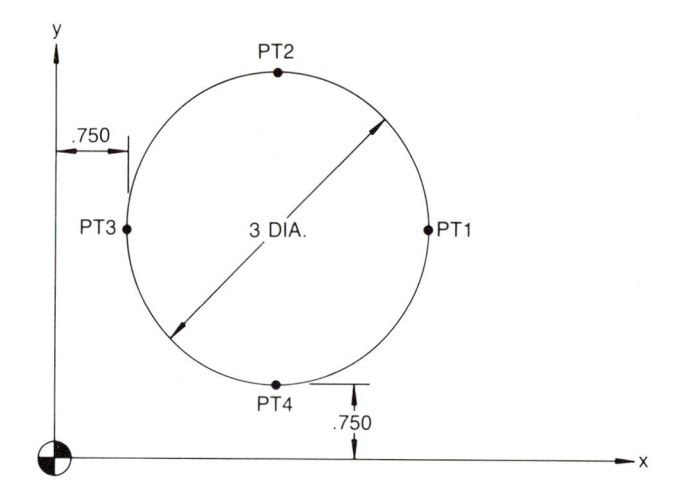

PROJECT TEN

Two Arcs in Full Quadrants

Circular interpolation is used to program any portion of a circle, such as arcs which occupy a full quadrant. All that has to be programmed are the coordinate locations of the start and end points of the arc, the radius of the arc, the coordinate location of the arc center, and the direction in which the cutter is to travel. This project involves the milling of the part boundary and the two arcs.

Notes

1. Material: ½-in.-thick mild steel (CS 100).

2. High-speed steel cutting tools are being used:
   (a) 1-in.-diameter, four-flute end mill (for sides).
   (b) ¼-in.-diameter, two-flute end mill (for circular slots).

3. Program clockwise in the absolute mode.

4. Include all speeds, feeds, and codes required to mill the sides and the slots.

5. Mill the slots 0.250 in. deep, starting at points 1 and 2.

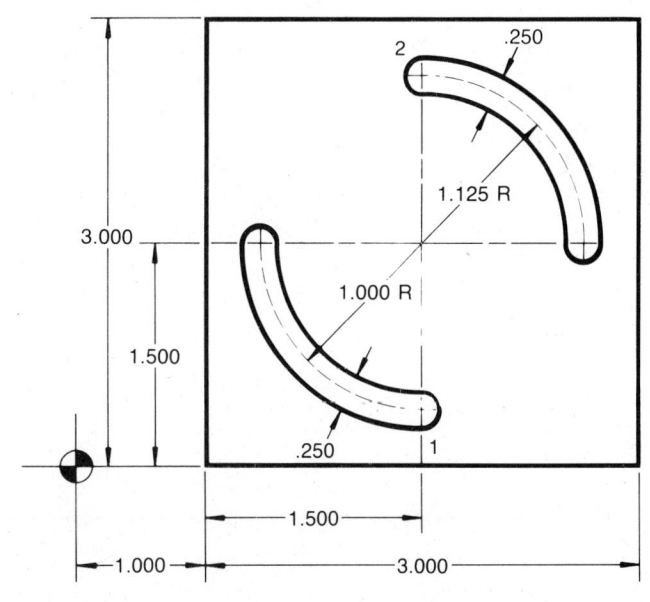

(The Superior Electric Company)

PROJECT ELEVEN

Circular Groove Milling

A full circle consists of four 90° arcs, and since many MCUs generate only one quadrant at a time, each arc must be programmed. Since the end point of one arc automatically becomes the start point of the next arc, a full circle requires the *first start point* and *four end points* to be programmed. This project involves milling a ⅜-in.-wide circular slot ¼ in. deep.

Notes

1. Material: mild steel (CS 100).
2. High-speed steel ⅜-in.-diameter, two-flute end mill.
3. Program *counterclockwise* in the absolute mode.
4. Include all speeds, feeds, and codes required to mill the circular groove.

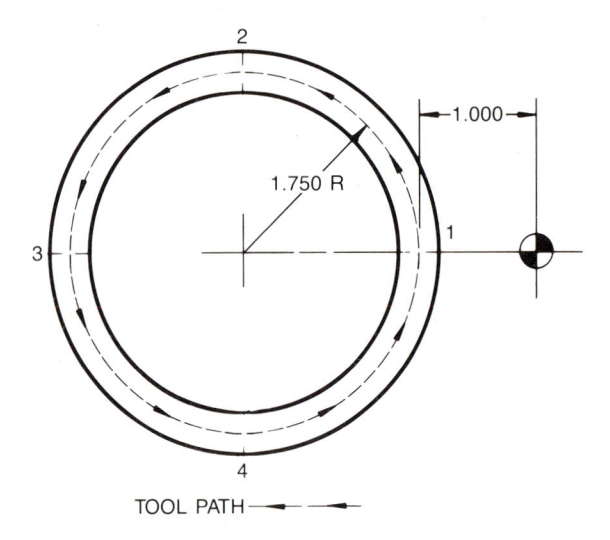

(The Superior Electric Company)

PROJECT TWELVE

A Full Circle Through Arcs

When programming a full circle or any portion of a circle by arcs or degrees, it is necessary to use trigonometry to calculate the start and end points of each arc. This project will improve the skills of the programmer by giving practice in calculating the XY locations of the start and end points of each arc or segment of a circle in order to generate the cutter path.

Notes

1. Calculate the XY locations of all points.
2. Start at XY zero.
3. Program the cutter path *counterclockwise* in the incremental mode.
4. Return to XY zero.

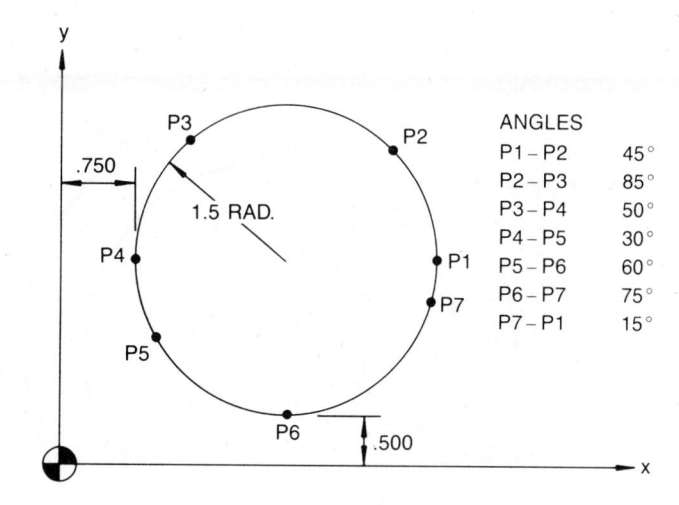

ANGLES	
P1 – P2	45°
P2 – P3	85°
P3 – P4	50°
P4 – P5	30°
P5 – P6	60°
P6 – P7	75°
P7 – P1	15°

1.5 RAD.

.750

.500

PROJECT THIRTEEN

Bolt Circle Drilling

Holes can be drilled on a bolt circle by mounting the workpiece on a rotary table and indexing the hole locations through the rotary table movements. If a rotary table is not available, or if the size or shape of the workpiece is not suitable for mounting on a rotary table, the coordinate (XY) location of all holes must be calculated. The holes can then be drilled by locating the machine table at the proper location for each hole. The operation consists of drilling four ¼-in.-diameter holes.

Notes

1. Material: ⅜-in.-thick steel (CS 125).
2. Calculate the XY locations of holes 2 and 3.
3. Include speeds, feeds, and codes required.
4. Program the holes in the absolute mode.

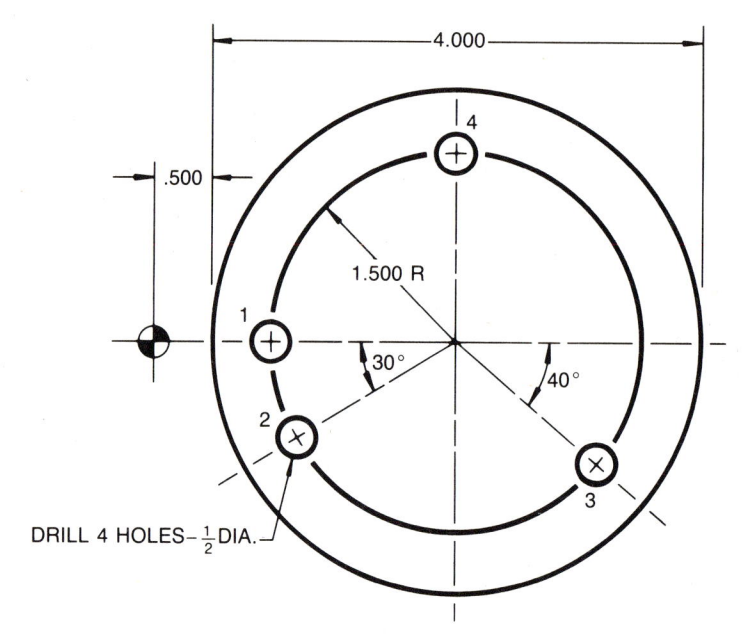

(The Superior Electric Company)

PROJECT FOURTEEN

Partial Circular Slot

Some circular slots are not cut for a full quadrant, or may extend into the next quadrant, and are dimensioned only as angles from the X and Y axis. In such a case, it will be necessary to calculate the XY locations of the *start* and the *finish* of the slot so that it can be programmed.

The part boundary and the circular slot must be machined.

Notes

1. Material: ¾-in.-thick mild steel (CS 100).
2. Cutting tools: high-speed steel:
   (a) 1-in.-diameter, four-flute end mill (for sides).
   (b) ¼-in.-diameter, two-flute end mill (for slot).
3. Calculate the start and end points of the arc or slot.
4. Program clockwise in the absolute mode.
5. Include all speeds, feeds, and codes required.

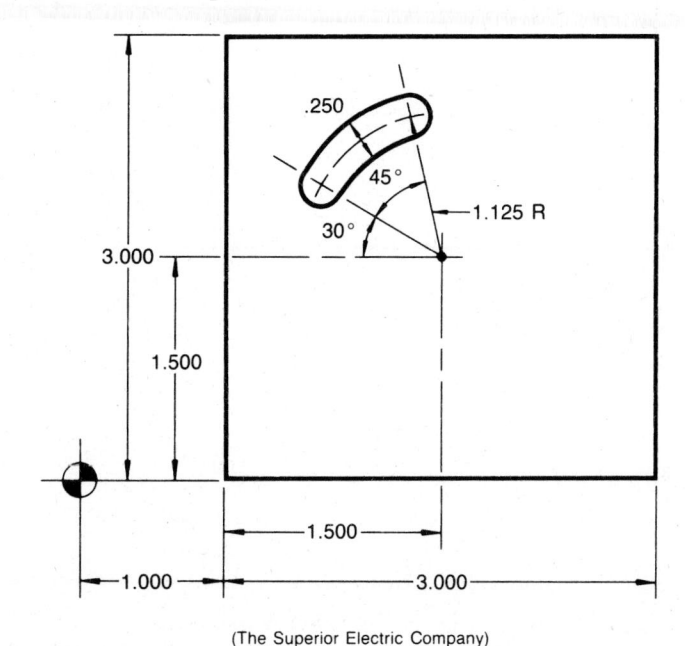

(The Superior Electric Company)

PROJECT FIFTEEN

Cutter Radius Tangency

Before starting to program any angular surface, it is necessary to calculate the required offset along the X or Y axis to find the amount that the cutter must be offset to produce the required angular surface. This project involves the programming of the cutter path around the workpiece. Cutter radius tangencies will have to be calculated in order to machine the part to the proper measurement.

Notes

1. Material: aluminum ¾ in. thick.
2. Cutting tool: high-speed steel four-flute end mill, ¾ in. diameter.
3. Program clockwise using absolute programming.
4. Calculate the cutter radius tangencies required to machine this part.

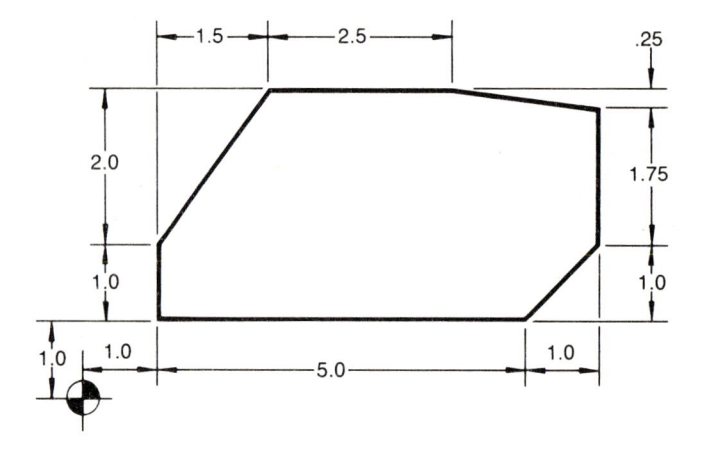

PROJECT SIXTEEN

Contouring Exercise

Circular interpolation can be used to generate arcs, curves, or free-form shapes which can be described with a series of arcs or circles. The programmer must program the start and end points of the arc, the XY location of the circle center, the radius of the arc, and the direction that the cutter is to travel. This project involves both linear and circular interpolation to program the cutter path for the workpiece.

Notes

1. Material: mild steel (CS 100); depth of cut 0.25 in.
2. Cutting tool: 1-in.-diameter high-speed steel end mill.
3. Program the part clockwise in the absolute mode.
4. Include speeds, feeds, and all codes required to machine this part.
5. Use coolant when machining.

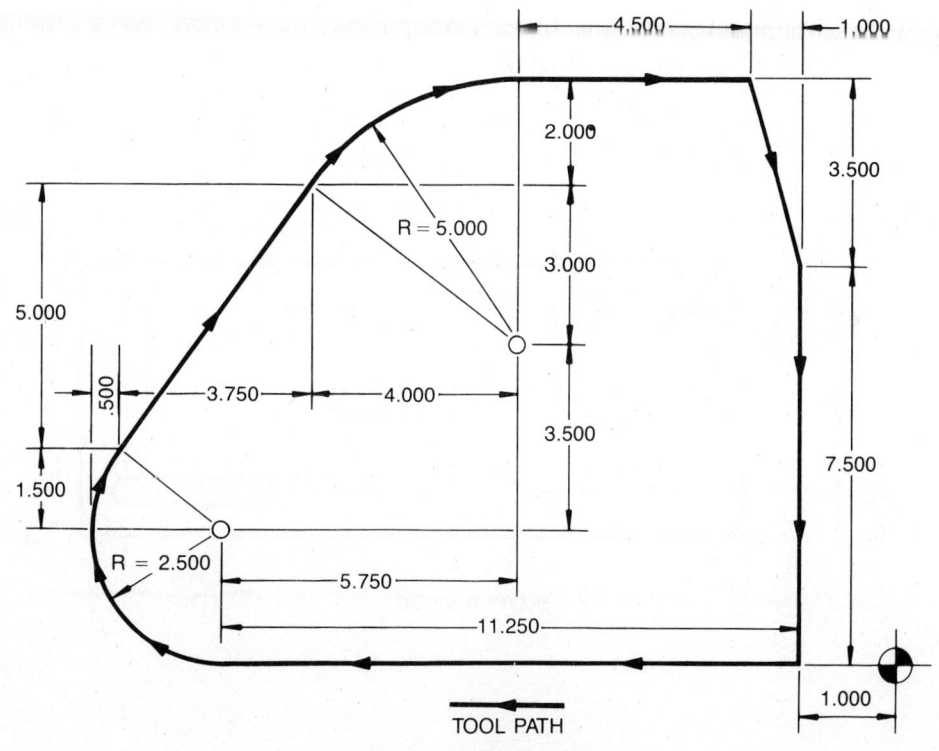

TOOL PATH

(The Superior Electric Company)

PROJECT SEVENTEEN

Macro

This project involves the programming of the part using macro programming. The macro program contains the information required to mill three similar slots, while the main program contains the information required to locate the cutting tool in position. Whenever a slot must be cut, the macro can be recalled into the main program. This feature enables data to be stored as a numbered subprogram for use in a main program when a repetitive pattern is required. Macro programming is usually written in the incremental format.

Notes

1. Material: mild steel.
2. Cutting tool: ½-in.-diameter, two-flute high-speed steel end mill.
3. Slot depth: 0.500 in.
4. Main program in absolute mode.
5. Macro program in incremental mode.
6. The part is located on the table X10.00 and Y8.00 in. from the machine home position.
7. Use the G92 code to start the program.

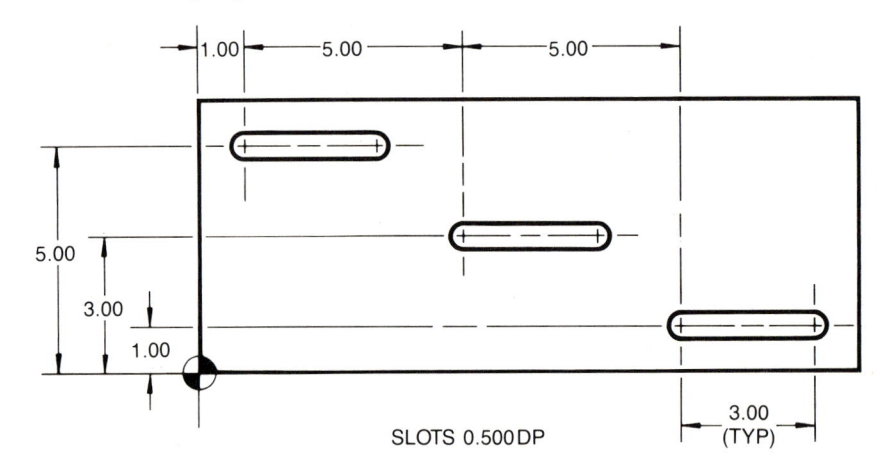

SLOTS 0.500 DP

PROJECT EIGHTEEN

Alphabet

The alphabet project is designed to provide the student with practice in linear interpolation.

Notes

1. Use incremental programming to program each letter.
2. Letters M and W are ¾ in. wide; all other letters are ½ in. wide.
3. The letters are ¾ in. high with a ⅛-in. spacing between each letter.
4. Plot the program to check the accuracy of each letter.

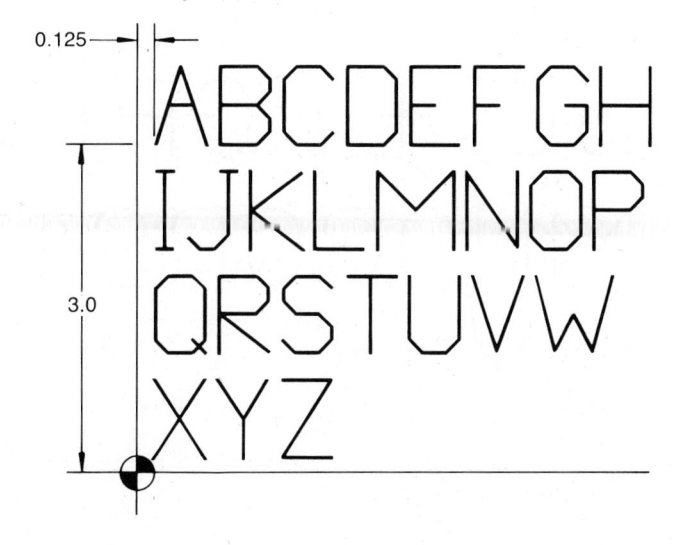

PROJECT NINETEEN

Circular Alphabet

This project involves the programming of the alphabet using both linear and circular interpolation.

Notes

1. Use absolute programming to program each letter.
2. Letters M and W are ¾ in. wide; all other letters are ½ in. wide.
3. The letters are ¾ in. high with a ¼-in. spacing between each letter.
4. Plot the program to check the accuracy of each letter.

PROJECT TWENTY

Cribbage Board

This project incorporates a number of the previous programming exercises. This program can be written using fixed drilling cycles, macro programming, or a more simplified method of programming the hole locations and board markings.

Notes

1. Material: wood, plastic, aluminum, brass, etc.
2. Cutting tools: high-speed steel 60° V-point cutter, 1/8-in.-diameter stub drill.
3. Use absolute mode for programming.
4. Program the board markings first.
5. Use G81 fixed drilling cycle to drill the holes.
6. Include all speeds, feeds, and codes required to machine the part.

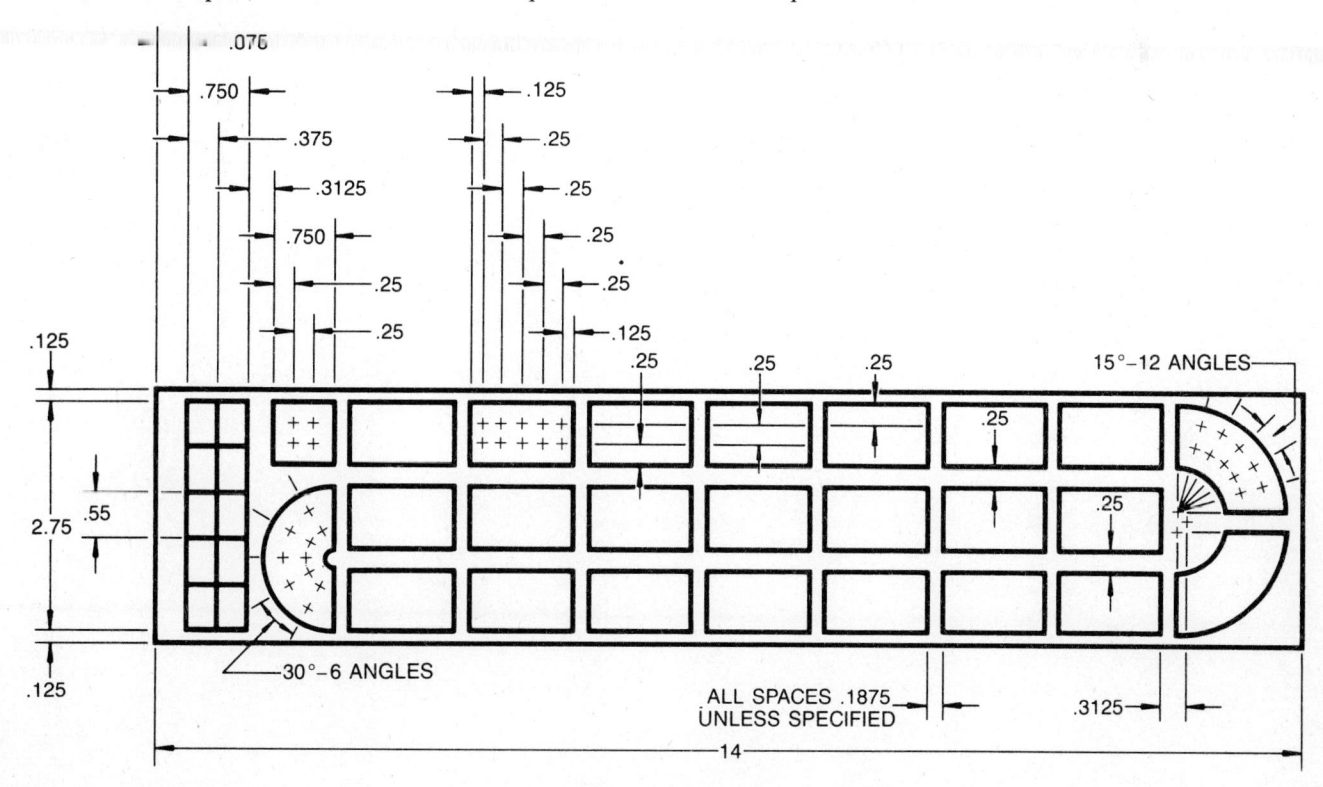

ALL HOLES .125 DIA. .5 DEEP

NC Wire EDM

PROJECT ONE

Thread Center Gage

This project involves linear interpolation consisting of straight lines and angular surfaces to program the part. Also involved is the setting of the parameters of the wire EDM machine to control the desired cutting conditions.

Notes

1. Material: stainless steel 1/16 in. thick.
2. Wire: 0.010-in.-diameter brass wire electrode.
3. Use the incremental mode, and program the part boundary clockwise.
4. Scale the drawing for sizes.
5. Use wire offset to compensate for size.
6. Complete the Wire EDM Cutting Conditions form provided by your instructor.

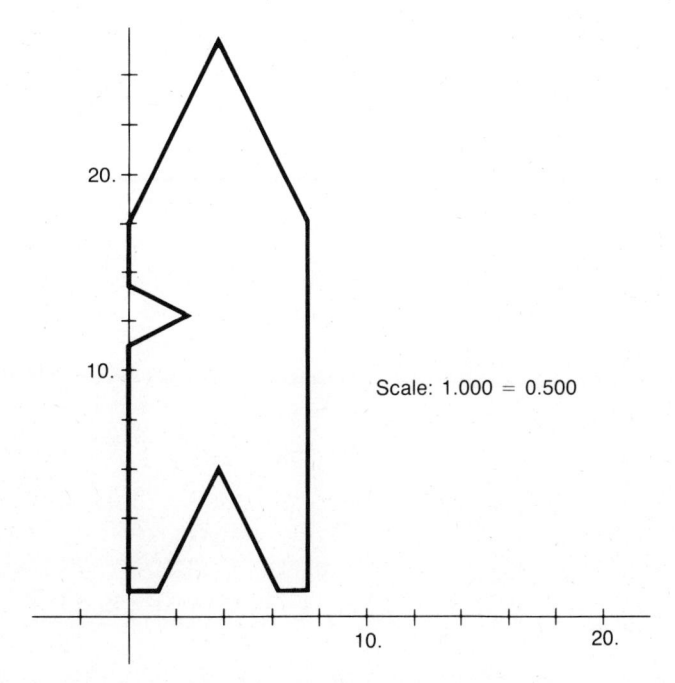

Scale: 1.000 = 0.500

Copyright © 1990 by McGraw-Hill, Inc. All rights reserved.

PROJECTS TWO, THREE, FOUR, FIVE, AND SIX
Shapes 1–5

These projects involve linear and circular interpolation to program the part boundary. These exercises will improve the programming skills of the programmer by giving practice in calculating the XY locations of the start and end points of each arc or segment of a circle in order to generate the part boundary.

Notes

1. Material: stainless steel ⅛ in. thick.
2. Wire: 0.010-in.-diameter brass wire electrode.
3. Use the absolute programming mode, and program the part boundary.
4. Complete the Wire EDM Cutting Conditions form.
5. Use wire diameter offset to compensate for size.
6. Scale drawings for sizes.

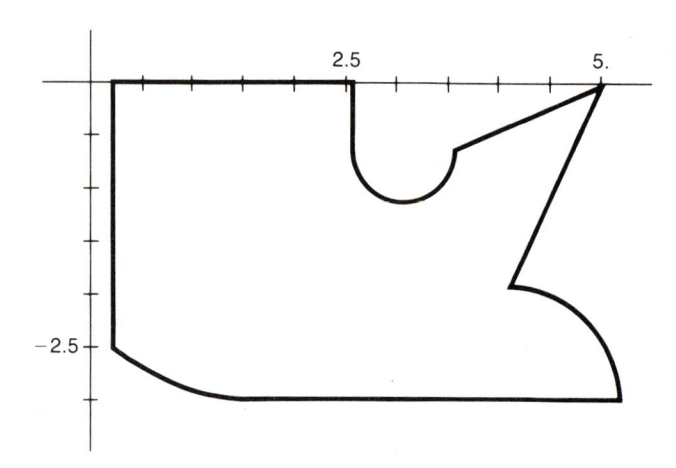

Shape 1

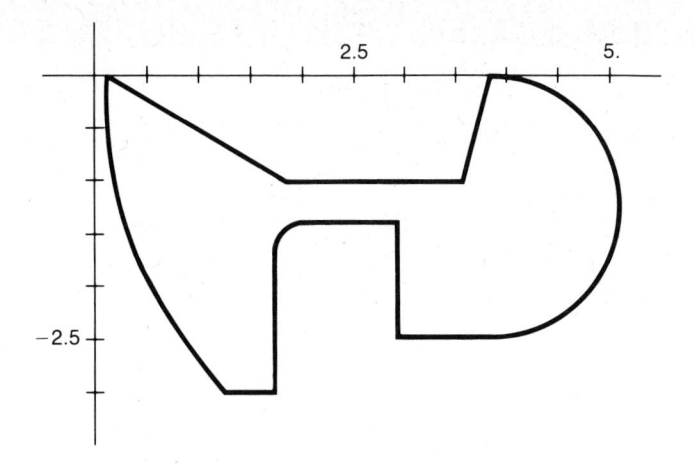

Shape 2

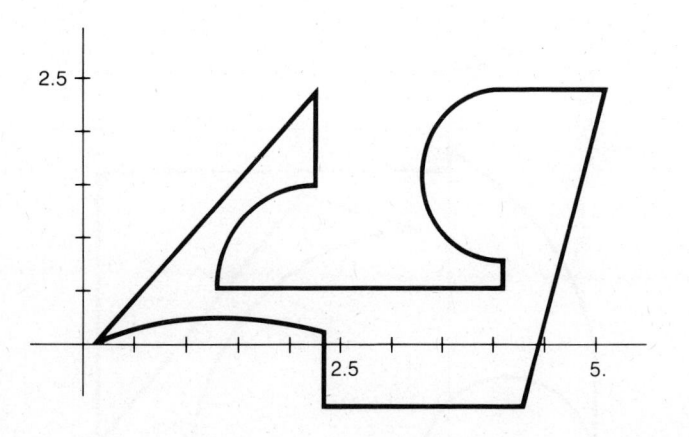

Shape 3

Copyright © 1990 by McGraw-Hill, Inc. All rights reserved.   **NC Wire EDM** PROJECTS TWO, THREE, FOUR, FIVE, AND SIX   71

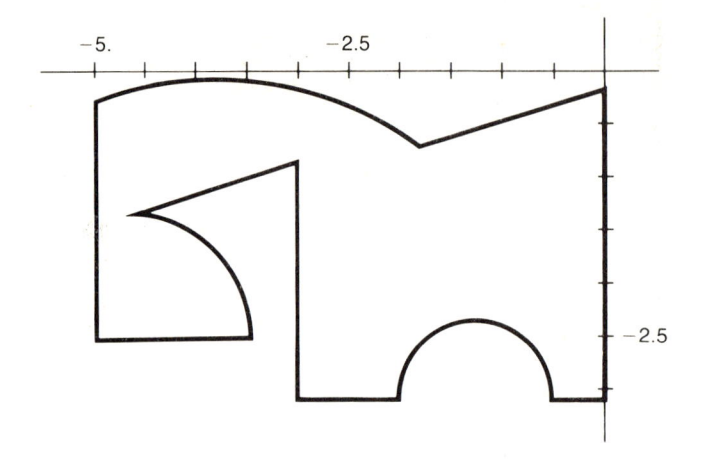

Shape 4

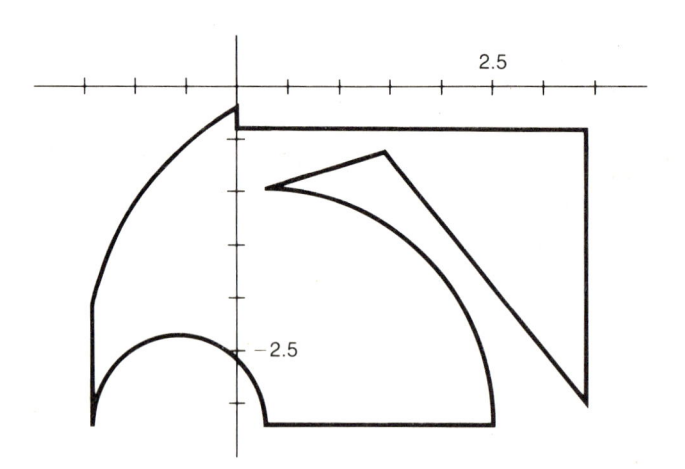

Shape 5

72 PROJECTS TWO, THREE, FOUR, FIVE, AND SIX   **NC Wire EDM**   Copyright © 1990 by McGraw-Hill, Inc. All rights reserved.

PROJECT SEVEN

C-Clamp

The clamp project includes linear and circular interpolation to complete the part boundary. Since the thread on the clamp screw is really a repetition of similar forms, a macro program could have been used which would have reduced the size of the program. The thread form in this program did not use a macro; therefore, each form was programmed individually.

Notes

1. Material: stainless steel ⅛ in. thick.
2. Wire: 0.008-in.-diameter brass wire electrode.
3. Use the absolute programming mode.
4. Program the part boundary *counterclockwise*.
5. Complete the Wire EDM Cutting Conditions form.
6. Scale the drawing for sizes.

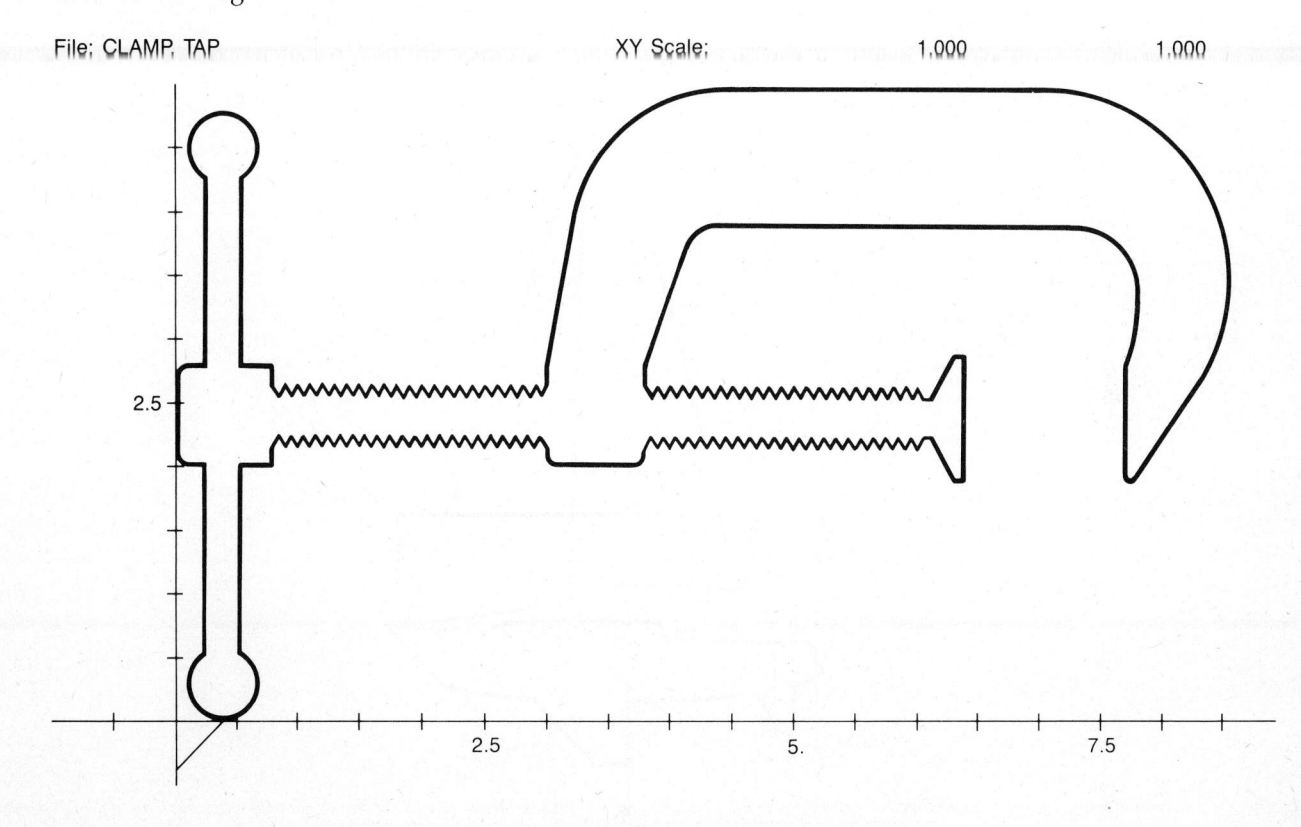

File: CLAMP.TAP   XY Scale: 1.000   1.000

PROJECT EIGHT

Sailboat

This project further develops the design skill of the programmer. Each student should be encouraged to design and program a sailboat of his or her own choice or a similar project which involves more complex linear and contour programming.

Notes

1. Material: brass plate 1/16 in. thick.
2. Wire: 0.008-in.-diameter brass wire electrode.
3. Use the incremental programming mode.
4. Program the part boundary *clockwise*.
5. Complete the Wire EDM Cutting Conditions form.
6. Scale the drawing for sizes.

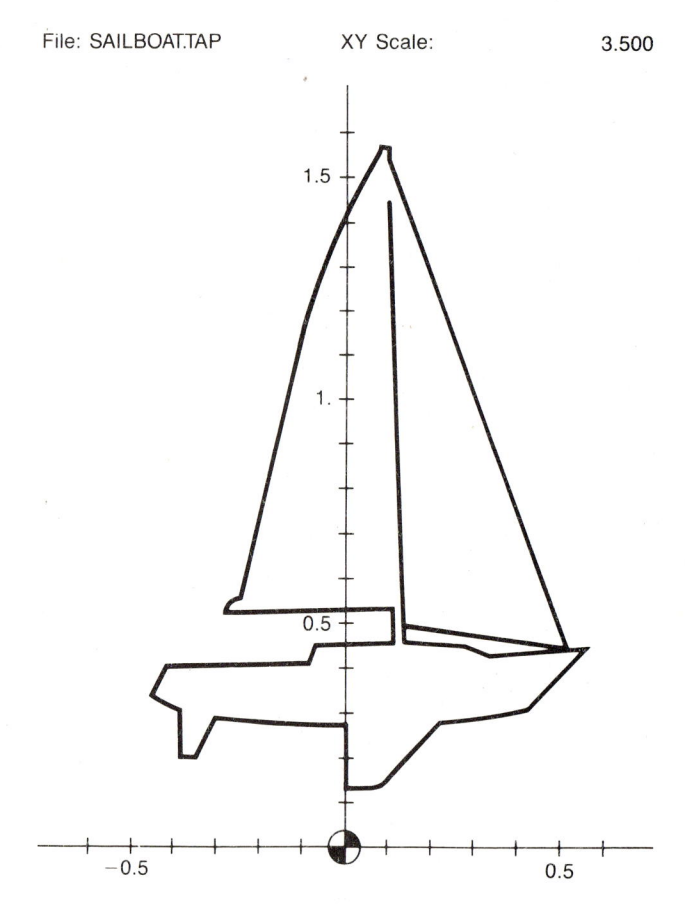

File: SAILBOAT.TAP   XY Scale: 3.500

PROJECT NINE

Witch

This project involves incremental programming and will improve the programming skills of the programmer. This program can also be produced using a digitizer tablet. This enables the programmer to produce an NC tape when there are no dimensions available on the drawing.

Notes

1. Material: stainless steel ¹⁄₁₆ in. thick.
2. Wire: 0.010-in.-diameter brass wire electrode.
3. Use incremental programming or a digitizer tablet.
4. Program the part boundary *clockwise*.
5. Complete the Wire EDM Cutting Conditions form.
6. Scale the drawing for sizes.

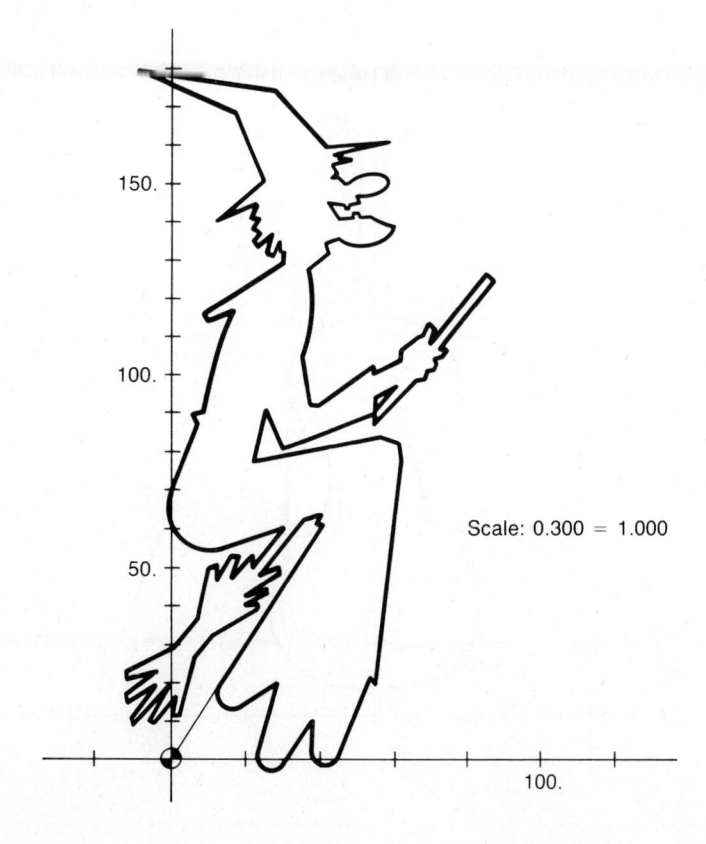

Scale: 0.300 = 1.000

Copyright © 1990 by McGraw-Hill, Inc. All rights reserved.

NC Wire EDM

PROJECT TEN

Violin

This project involves the use of a digitizer tablet to program the required part boundary. The project can be programmed at the size shown or can be scaled up or down before programming.

Notes

1. Material: brass plate ¼ in. thick.
2. Wire: 0.010-in.-diameter brass wire electrode.
3. Digitize the part boundary *counterclockwise*.
4. Produce an NC tape.
5. Complete the Wire EDM Cutting Conditions form.

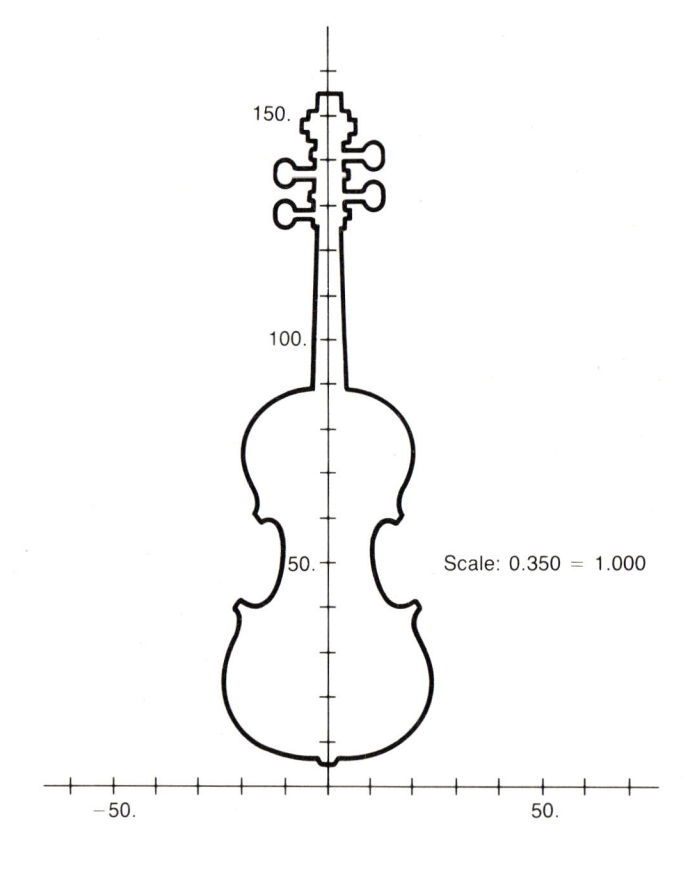

Scale: 0.350 = 1.000

NC Turning Center

PROJECT ONE

Two-Step Locating Pin

This project consists of machining two parallel diameters with the workpiece held in a chuck. This provides practice in linear interpolation where the program is along the X axis (tool infeed) and the Z axis (longitudinal feed). It involves the use of rapid feed, setting depths of cuts, and tool retract.

Notes

1. Material: aluminum 1 in. diameter (CS 500).
2. Cutting tool: diamond-shaped carbide.
3. Program in the absolute mode.
4. Feed rate 1 in./min.
5. Spindle rotation counterclockwise (CCW).
6. Part zero as indicated on the drawing.

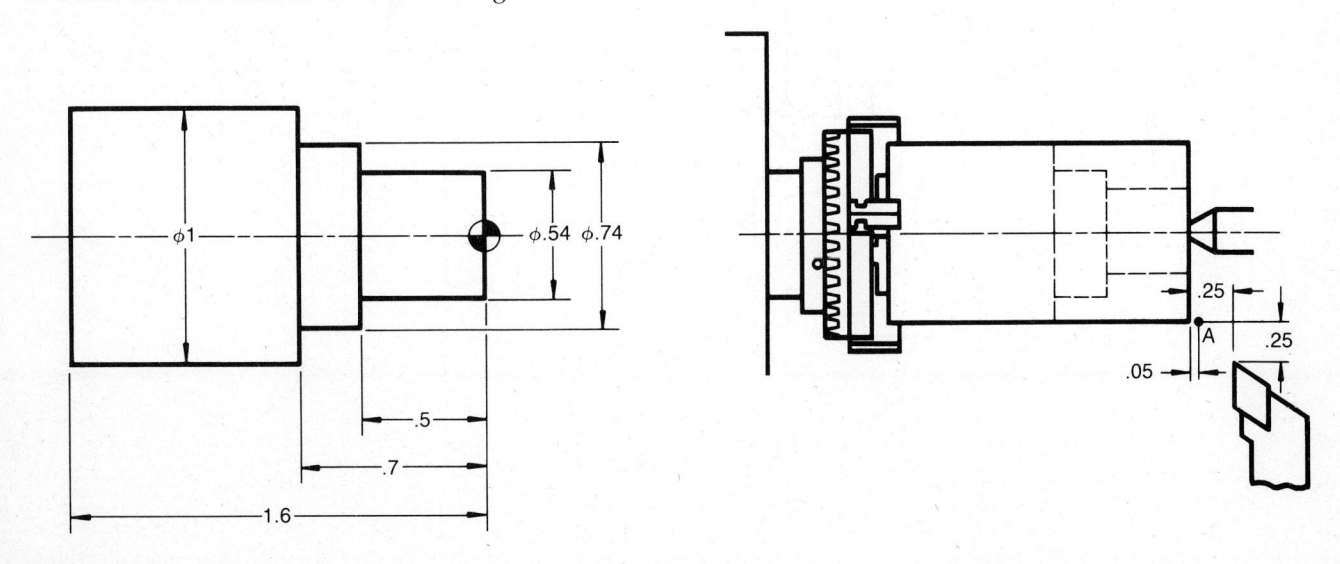

(Emco-Maier Corporation)

PROJECT TWO

Taper Shaft

This project consists of machining two parallel diameters and a taper. It involves the use of linear interpolation to cut the taper on the shaft. To program any taper, the start point and the end point of the taper must be calculated and programmed. The workpiece should be held in a chuck for machining.

Notes

1. Material: aluminum 1 in. diameter (CS 500).
2. Cutting tool: diamond-shaped carbide.
3. Program in the absolute mode.
4. Feed rate 0.010 in./rev.
5. Spindle rotation counterclockwise (CCW).
6. Part zero as indicated on the drawing.

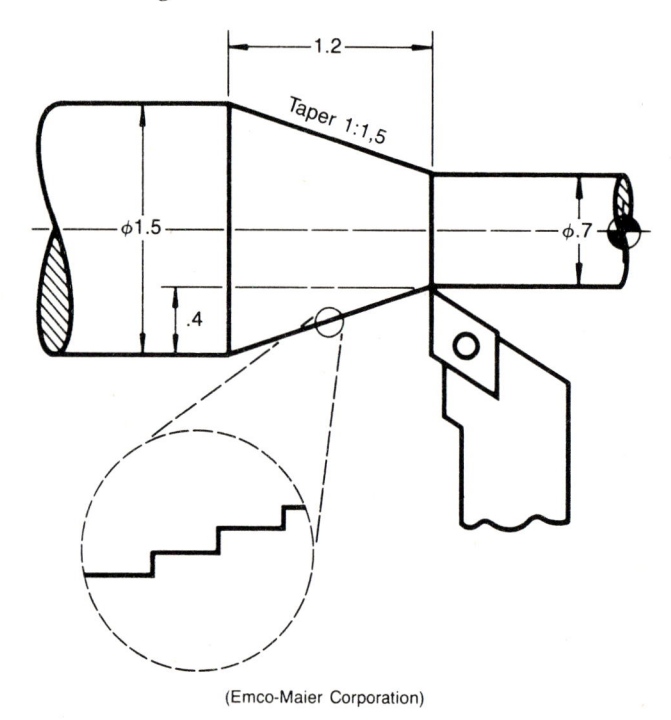

(Emco-Maier Corporation)

PROJECT THREE

Spindle

This project includes linear and circular interpolation in order to program the part boundary to produce the required form. It involves setting various tool depths to cut the tapers, parallel diameters, chamfers, and radius. Tool change positions are required to index the drill into position to produce the ½-in.-diameter hole. The workpiece should be held in a three- or four-jaw chuck for machining.

Notes

1. Material: aluminum.
2. Cutting tool: diamond-shaped carbide, ½-in.-diameter stub drill.
3. Program in the absolute mode.
4. Spindle rotation counterclockwise (CCW).
5. Part zero as indicated on the drawing.

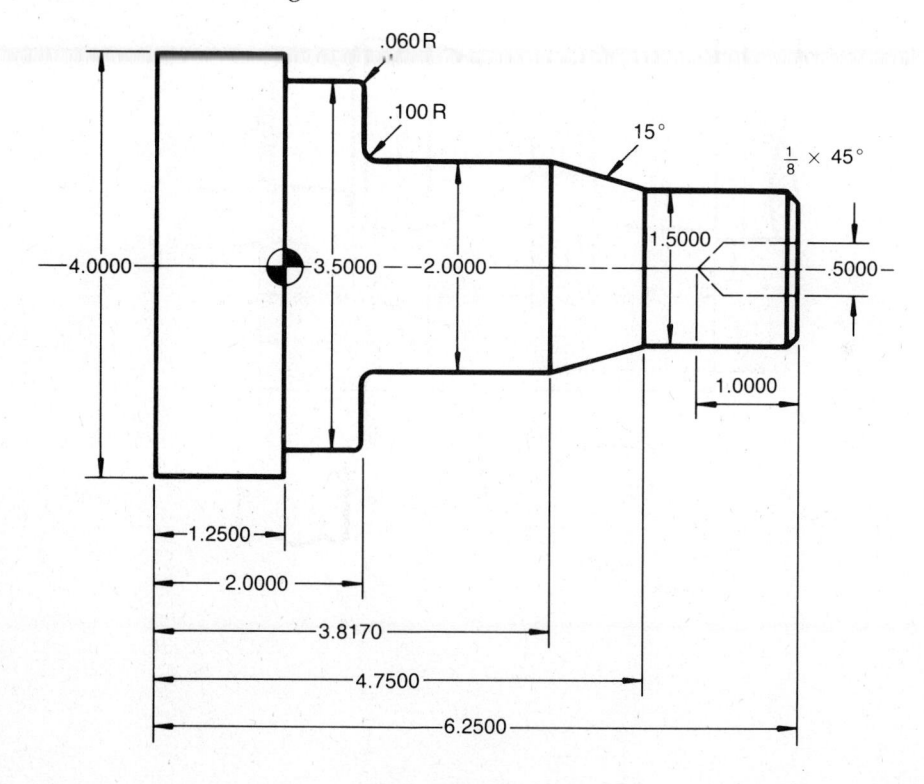

(Cincinnati Milacron Incorporated)

PROJECT FOUR

Contour Form

This project further develops the programmer's skills in linear and circular interpolation. It involves the machining of parallel diameters, taper, radius, and chamfers to produce the part. The workpiece should be held in a chuck for machining purposes.

Notes

1. Material: aluminum.
2. Cutting tool: diamond-shaped carbide.
3. Program in the absolute mode.
4. Spindle rotation counterclockwise (CCW).
5. Part zero as indicated on the drawing.

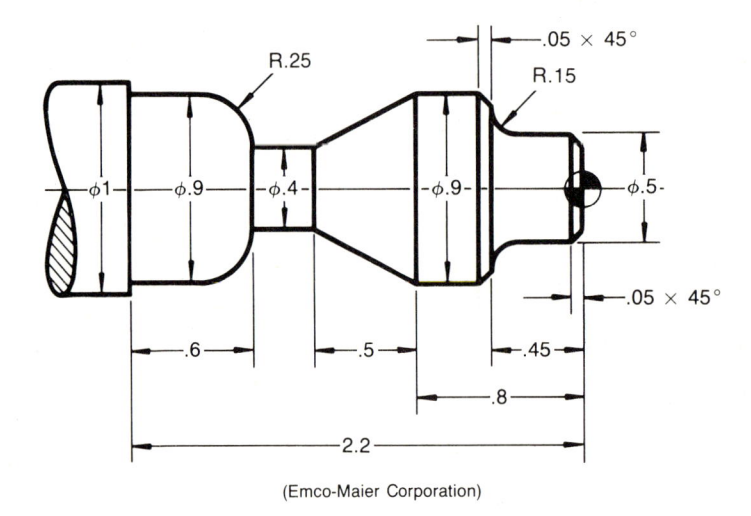

(Emco-Maier Corporation)

80 PROJECT FOUR NC Turning Center   Copyright © 1990 by McGraw-Hill, Inc. All rights reserve

PROJECT FIVE

Pawn

The pawn, one of the pieces in a chess set, will further develop the skills of a programmer. It involves linear and circular interpolation for producing parallel diameters, tapers, contours, and chamfers. Most work machined in a lathe involves these common operations, and programming this part of the set will provide valuable experience.

Notes

1. Material: aluminum, brass, plastic, etc.
2. Cutting tool: diamond-shaped carbide.
3. Program in the absolute mode.
4. Spindle rotation: counterclockwise (front turret).
5. Part zero as indicated on the drawing.

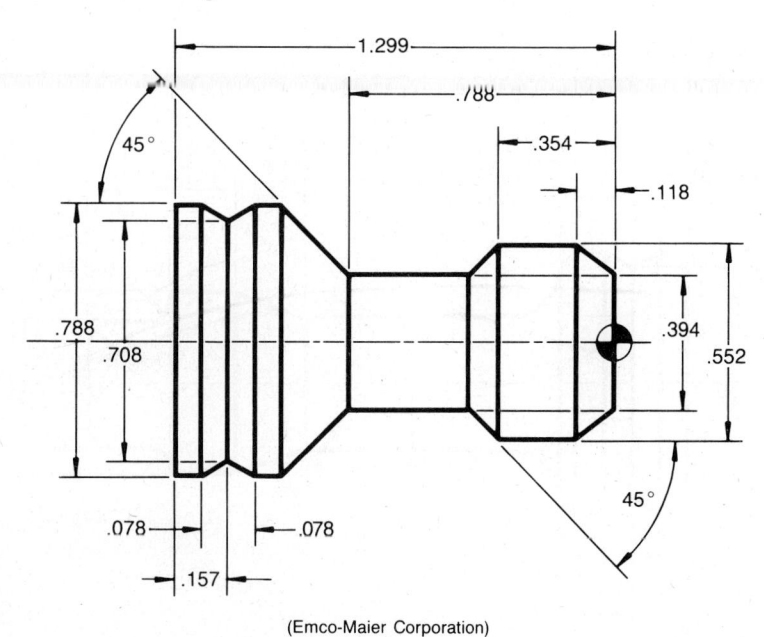

(Emco-Maier Corporation)

NC Turning Center PROJECT FIVE 81

PROJECT SIX

Bishop

The bishop, one of the pieces in a chess set, will further develop the skills of a programmer. It involves linear and circular interpolation for producing parallel diameters, tapers, contours, and chamfers. Most work machined in a lathe involves these common operations, and programming this part of the set will provide valuable experience.

Notes

1. Material: aluminum, brass, plastic, etc.
2. Cutting tool: diamond-shaped carbide.
3. Program in the absolute mode.
4. Spindle rotation: clockwise (rear turret).
5. Part zero as indicated on the drawing.

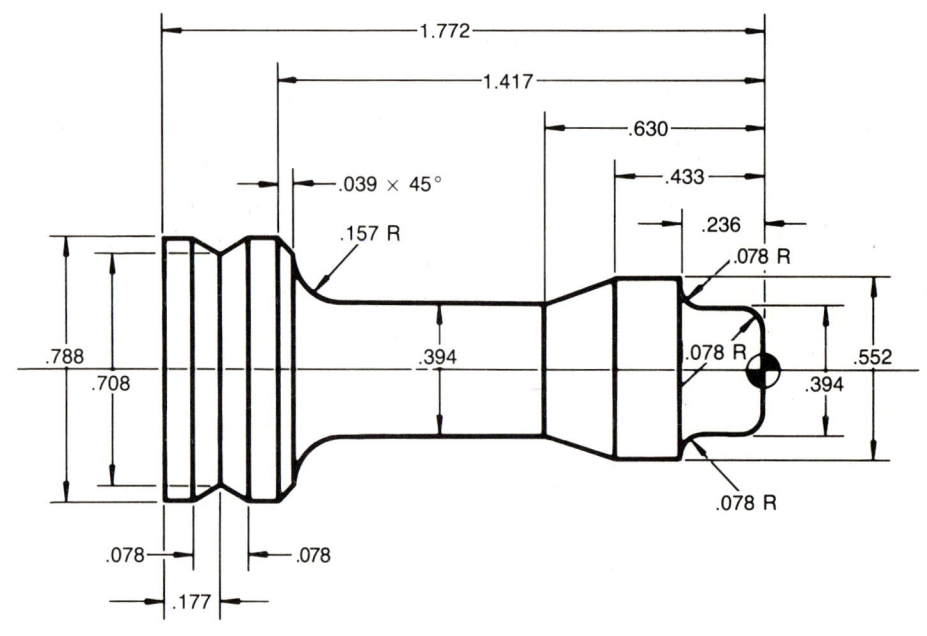

(Emco-Maier Corporation)

82 PROJECT SIX NC Turning Center

PROJECT SEVEN

Castle

The castle, one of the pieces in a chess set, will further develop the skills of a programmer. It involves linear and circular interpolation for producing parallel diameters, tapers, contours, and chamfers. Most work machined in a lathe involves these common operations, and programming this part of the set will provide valuable experience.

Notes

1. Material: aluminum, brass, plastic, etc.
2. Cutting tool: diamond-shaped carbide.
3. Program in the absolute mode.
4. Spindle rotation: clockwise (rear turret).
5. Part zero as indicated on the drawing.

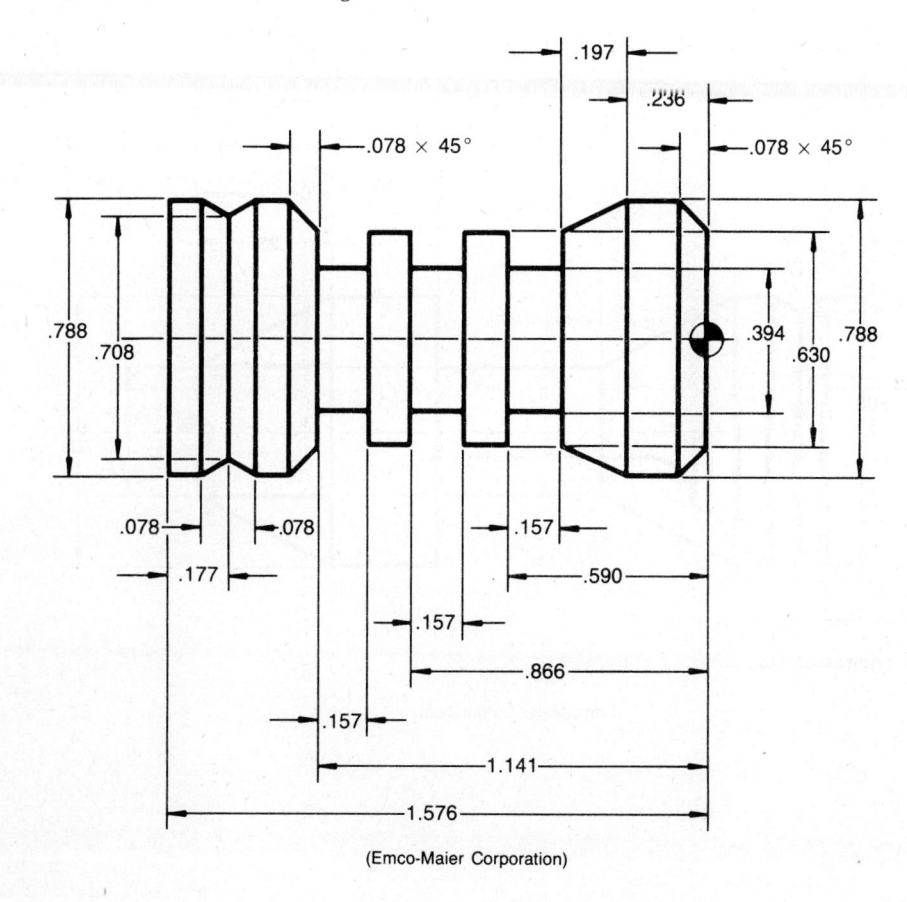

(Emco-Maier Corporation)

PROJECT EIGHT

Knight

The knight, one of the pieces in a chess set, will further develop the skills of a programmer. It involves linear and circular interpolation for producing parallel diameters, tapers, contours, and chamfers. Most work machined in a lathe involves these common operations, and programming this part of the set will provide valuable experience.

Notes

1. Material: aluminum, brass, plastic, etc.
2. Cutting tool: diamond-shaped carbide.
3. Program in the absolute mode.
4. Spindle rotation: clockwise (rear turret).
5. Part zero as indicated on the drawing.

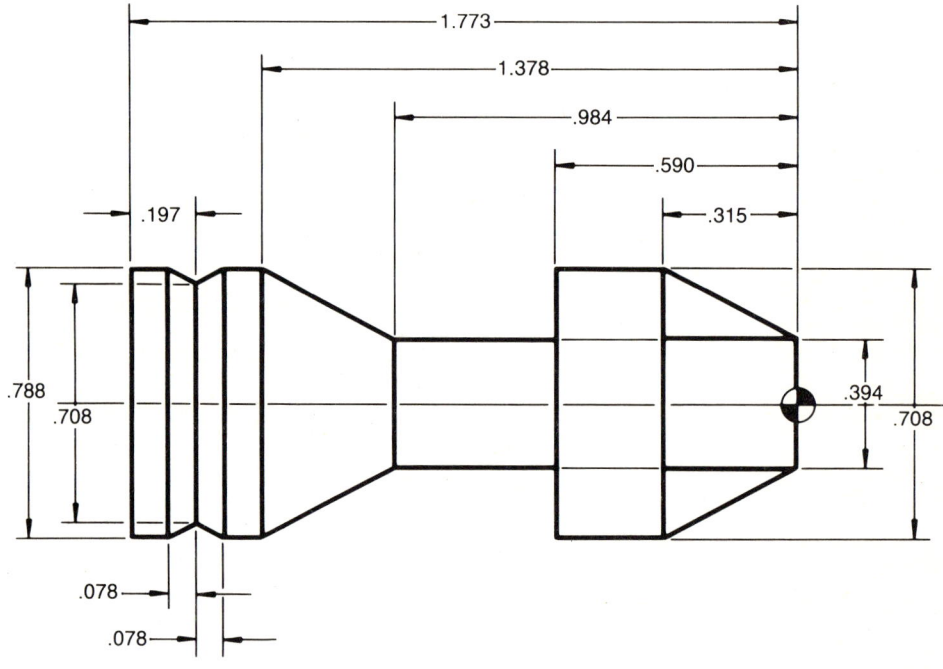

(Emco-Maier Corporation)

84 PROJECT EIGHT NC Turning Center

PROJECT NINE

Queen

The queen, one of the pieces in a chess set, will further develop the skills of a programmer. It involves linear and circular interpolation for producing parallel diameters, tapers, contours, and chamfers. Most work machined in a lathe involves these common operations, and programming this part of the set will provide valuable experience.

Notes

1. Material: aluminum, brass, plastic, etc.
2. Cutting tool: diamond-shaped carbide.
3. Program in the absolute mode.
4. Spindle rotation: clockwise (rear turret).
5. Part zero as indicated on the drawing.

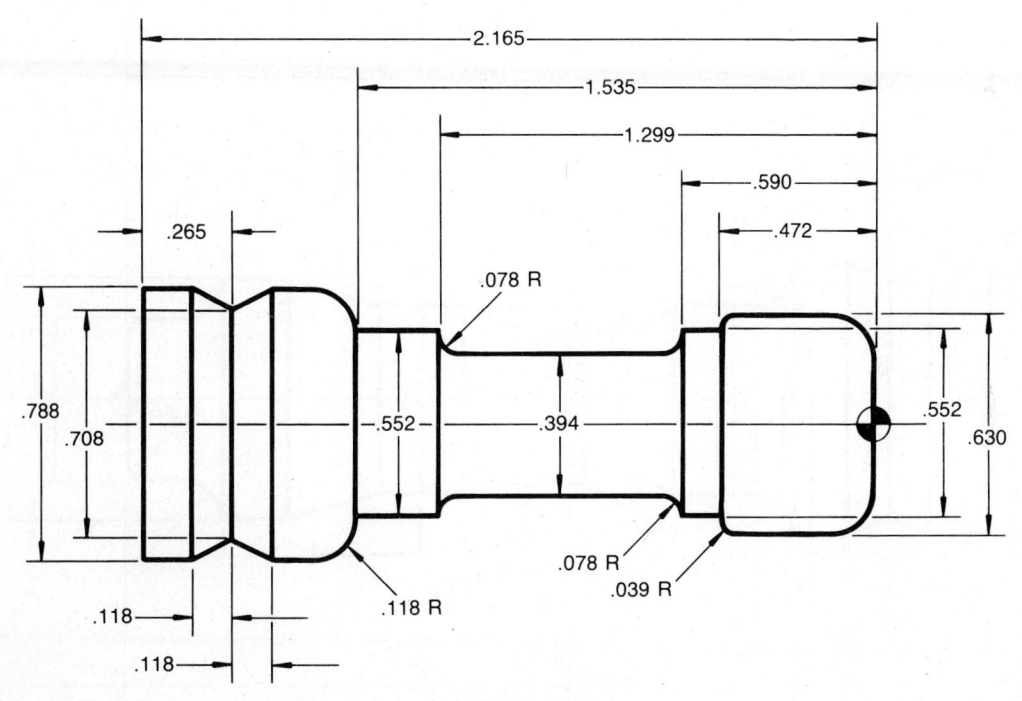

(Emco-Maier Corporation)

NC Turning Center PROJECT NINE 85

PROJECT TEN

King

The king, one of the pieces in a chess set, will further develop the skills of a programmer. It involves linear and circular interpolation for producing parallel diameters, tapers, contours, and chamfers. Most work machined in a lathe involves these common operations, and programming this part of the set will provide valuable experience.

Notes

1. Material: aluminum, brass, plastic, etc.
2. Cutting tool: diamond-shaped carbide.
3. Program in the absolute mode.
4. Spindle rotation: clockwise (rear turret).
5. Part zero as indicated on the drawing.

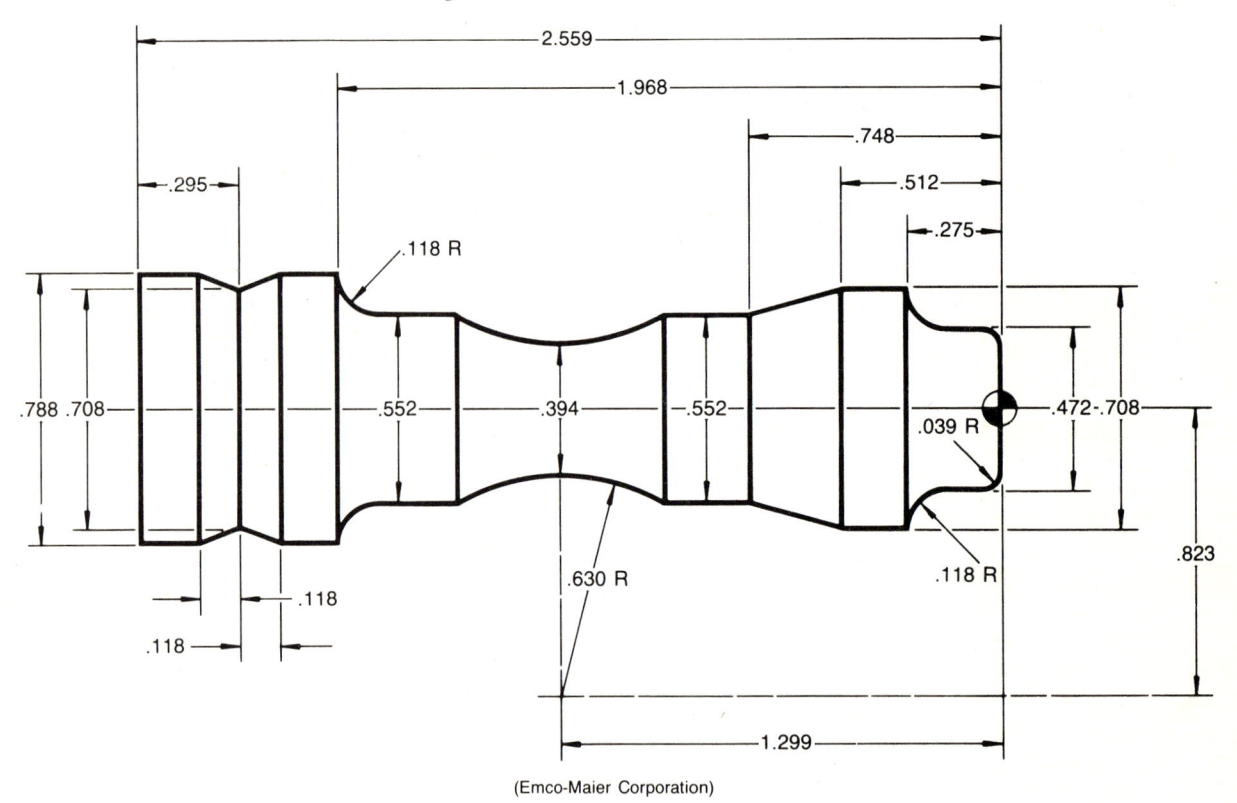

(Emco-Maier Corporation)